How to Build Military Grade Suppressors

.22 thru .50 BMG

Desert Publications

El Dorado, AR 71730 U. S. A.

How to Build Military Grade Suppressors .22 thru .50 BMG

Published by Desert Publications
215 S. Washington Ave.
El Dorado, AR 71730 U.S.A.
1-870-862-3811
info@deltapress.com

Printed in U.S.A
ISBN: 978-0-87947-195-8
10 9 8 7 6 5 4 3

Desert Publication is a division of
The DELTA GROUP, Ltd.

Direct all inquiries & orders to the above address.

Neither the author nor the publisher assumes any responsibility for the use or misuse of the information contained in this book. This material was compiled for educational and entertainment purposes and one should not construe that any other purpose is suggested.

Warning!

The Publisher (Desert Publications) produces this book for informational and entertainment purposes and under no circumstances advises, encourages or approves of use of this material in any manner.

Contents

How to Build Military Grade Suppressors

Introduction

My first book, "How To Build Practical Firearm Suppressors," was primarily about improvised suppressors, of the type which can be made from common hardware store items, with little expense and no expertise. There are a number of books with designs in this category. They all have several things in common. They're cheap, easy to build, and fun to play with. Very fun. It is an uncommon, (and very dull,) breed of man who can fire off a clip from an effectively suppressed weapon without bouncing up and down giggling. Another thing most of them have in common, mine included, is that you wouldn't really want one with you on a combat mission. Improvised suppressors just cannot be built to combat quality standards. They break when banged around, they blow up, the baffles wear out quickly, and most of them destroy accuracy.

The purpose of this book is to explain procedures for building genuine, combat quality suppressors for military type arms. If you just want to make something cheap to play around with on the weekends in a short amount of time, then pick up a copy of my first book. Real quality does not come for under ten dollars, or without any skills to speak of. This book is for people who want the real deal. The designs presented here, while being simple to construct, make use of an advanced baffle design that is optimized for this construction method. They compete quite nicely with commercial suppressors in terms of effectiveness, durability, and size required for the unit.

On the subject of noise level testing, I regret to say that I never got to use the proper sound testing equipment as described in the book "Silencer History and Performance."I searched far and wide for that piece of equipment, and finally found it for rent for five hundred bucks for two weeks. Sorry, but that's above the call of duty, and so at this time I can't give exact decibel reduction ratings for each model. What I can tell you is that I've shot a lot of different commercial suppressors for comparison. And when I was developing my own design, I tried every variation you can possibly imagine in baffle design, packing material, and so on, trying every idea I could think of that would work for this construction method.

For testing, I would take two suppressors to a clear area on a quiet night and fire them side-by-side. Firing in the air in an area clear of trees and hills is the necessary procedure to prevent sound from the muzzle or supersonic crack from being reflected back at you and interfering with your ability to hear just the sound of the muffled muzzle blast. In both rifle and pistol calibers, the results were very consistent, and the design presented in this book was the clear-cut winner.

At this point, let us review briefly just what a suppressor is and how it works. There are three primary sources of noise from firearms.The first is the muzzle blast, caused by rapidly expanding gases from the barrel of the gun. This is the noise that is acted upon by the suppressor. Essentially what a suppressor is, is a chamber, or series of chambers, designed to contain and muffle the expanding gases while allowing the bullet to pass through.

The second cause of noise is the supersonic crack of the bullet as it breaks the sound barrier. The speed of sound is approximately 1100 feet per second. In order to eliminate this noise, it is necessary to select a caliber or load that travels about 1050 feet per second or slower. There is also specially made subsonic ammo in many calibers, including rifle calibers, specially made to travel slower than sound.

The third cause of sound is the action of the gun. This is especially true of semi-automatic weapons. Normally this sound is drowned out by the muzzle blast, but when the blast is suppressed, the sound becomes more apparent. This noise is much smaller than the other noises though, so it is not generally a great concern.

Another noise that is less talked about is bullet impact. A heavy, large caliber bullet can make quite a smack, or splat, depending on what you're shooting. Keep that in mind when you think about sniping sentries with something in .45 caliber.

There is also an important distinction to be made between semi-auto weapons of pistol caliber and of rifle caliber, one that results in a bit of a dilemma. Pistol caliber weapons operate by blowback action, and rifle caliber weapons are gas-operated. A gas-operated rifle doesn't start to cycle until the muzzle blast pressures engage the cycling mechanism, by which time the chamber pressures have dropped. In a pistol caliber blowback action, chamber pressures are relatively high when the breech opens, allowing a substantial portion of the blast to escape through the breech and increasing the noise. Barrel length has a great effect on this. A barrel that is cut down to 5" or 6" will allow the pressure to drop before the case clears the chamber, reducing this problem greatly but not eliminating it entirely. A rifle caliber weapon, while not having the breech-blast problem, most often requires supersonic ammo to cycle the action. There are but a very few exceptions to this, such as the .300 Whisper caliber. The .300 Whisper is a .221 Fireball or .223 case necked up to a .30 caliber, and made to fire either a heavy 200 to 250 grain bullet at subsonic velocities, or lighter bullets at supersonic velocities. A properly made AR-15 in this caliber will cycle with both types of ammo, which solves this dilemma, the main downside being that the ammo costs a fortune and is not what you'd call a common caliber.

The designs and construction methods shown here are made especially for military arms. The construction of all the different models is pretty similar, and designs are included for a variety of weapons, including sniper rifles, assault rifles, submachine guns, and some pistols. These suppressors are built entirely of steel. There are no rubber baffles to wear out, no plastic parts to break. Their solid welded construction insures that they will never break when you need them, even with full auto fire.

Because these suppressors are designed for military arms, you will most likely need to select a weapon with a threaded barrel, because flash suppressors and muzzle brakes are the basis for construction and attachment. If your weapon isn't threaded, there are threading services available that can fix that problem for you. Just go on the internet and do a web search under "barrel threading services" if you need to find one. There are also threaded aftermarket barrels available for a variety of weapons, especially pistols.

You can find various other types of slip-on, pin-on, and other types of flash suppressors and muzzle brakes. They can suffice for .22's or some lighter pistol caliber weapons, but not for heavy rifle calibers. Those require absolute stability. So choose your weapon carefully, and don't go second best and figure on using an improvised attachment method that may or may not do the job.

How to Build Military Grade Suppressors

Construction Tools and Materials

CHAPTER 1

The following tools and machines are what you will need before beginning.

1. Welding machine. Welding machine types are discussed in the "Welding Equipment and Procedures" section.
2. Hand held drill, preferably with a 1/2" drive.
3. Dremel tool, 110v powered, with die grinder bit.
4. Bench grinder.
5. Hacksaw.

Certain other tools may be required, and are detailed in the steps for construction.

Outer Tubes: The outer tubes of the suppressors are made from chain link fence poles. These come in four different diameters. The inside diameters of the tubes are as follows: 30mm, 1 1/2", 1 3/4", and 2 1/4". They are labeled respectively as 1 3/8", 1 5/8", 1 7/8", and 2 3/8". The small 1 3/8" size isn't as easy to find, but is sold by the Home Depot chain.

Baffles: The baffles of the suppressor are made from automotive freeze plugs. These are available from auto parts stores, through the mail, or on the internet if you need special sizes or quantities. The size of plug you select will be the same as the inside diameter of the outer tube of the suppressor.

Diffusing Material: Stainless steel screen is used for the diffusing material, which is placed in between the baffles.

Aluminum window screen is not suitable for this purpose, except maybe for very small calibers like .22. It is too flimsy and will be blown out of place and into the center of the suppressor, ruining it permanently. Heavy stainless screen is more expensive and harder to find, but it is far more durable and will not have this problem. There are two main types of stainless screen; residential grade and industrial grade. Residential grade is even flimsier than aluminum screen and is entirely unsuitable. Industrial grade screen comes in several different mesh sizes and wire thicknesses, and the type needed is dependent on the caliber you are using. See the "Source Directory" section for a supplier and the sizes you'll be needing.

One of the reasons these suppressors compare so well to commercial models is because of the use of stainless screen for diffusing material. Most commercial suppressors just use baffles with no diffusing material. This is a shame because the screen has a huge impact on effectiveness. It is understandable why they don't use it though, because the screen gets clogged with fouling, requires cleaning more often, and shortens the life of the unit. That's not a problem if you're making your own though, because it's not like you have to shell out another $400 plus $200 tax to replace it. Just hit the shop and whip yourself up a new one. Well, that is unless you're one of the poor souls who have to concern themselves with paperwork issues. Even then, you can just place your serial number close to the rear end of the unit, and when it wears out, just cut it off forward of the number and replace it from there.

As the freeze plug baffles are welded in, the hole for the bullet is made in a unique way, by just shooting through it, often with specially made hole-punching ammo made for this purpose. The hole is then reamed with a drill bit to be larger than the bullet, to prevent the bullet and baffle from touching, and the baffle is also then formed into an "indented" design made to redirect muzzle blast across the path of the bullet. This construction method allows you to make the hole as small as possible while being perfectly aligned with the bullet path. With no rubber or plastic baffles touching the bullet, full accuracy is maintained. It sounds scary to punch the holes this way, I know. It took me quite some time to work up the nerve to try it, but I've now done it with several different types of guns with only one minor problem. That problem was with a Makarov pistol. The slide blew back hard enough to disengage it from the frame. I replaced it back on, and it worked fine again. Later, it happened a couple more times. Then I noticed that the top part of the trigger guard, which serves to stop the slide, was being dented and dinged by the impact of the slide. Now when using the Makarov or other semi-auto pistol to punch holes, I grip the slide with my left hand using heavy leather gloves to hold it closed when firing to prevent this from happening. Semi-auto pistols are most susceptible to damage caused by the excessive pressures, so this is the technique you should plan on using when using them to punch holes. The Beretta M-9 already has had slide failures resulting from shooting with suppressors and high-powered submachine gun ammo. Some of these slide failures caused serious injury, so be extra careful with these types of guns. Also, when using any type of weapon to punch holes, hold the gun away at an angle so a bolt or slide or other part of the gun could not blow back and nail you, just in case.

The use of special hole-punching ammo can minimize these problems. Read the section on hole-punching ammo for details. Always weld in and shoot through just one baffle at a time. After shooting, the hole has to be reamed and formed into the indented design before moving on to the next one. Also, bullet expansion will begin with the first baffle, making subsequent holes larger and uglier if you shoot through more than one.

The first edition of this book focused on older type designs that use a centerpipe vented into an expansion chamber. The more modern designs in this book use baffles only. I considered putting both types of designs into this edition, but the truth is that the baffle-only designs are so much more effective that there is really no point in learning to use centerpipe designs, which are completely obsolete. About all a centerpipe accomplishes is to direct muzzle blast straight down the center of the unit, greatly reducing effectiveness.

How to Build Military Grade Suppressors

Hole-Punching Ammunition

CHAPTER 2

Following are instructions on how to make or select special ammo for punching holes in baffles. The idea here is to make ammo that will punch a good, properly formed hole, while being quiet enough you don't have to drive way out in the hills to punch every hole. These instructions require no reloading equipment which works out very well since I don't have any.

.22 Long Rifle: For .22 long rifle, simply use CCI Stinger ammunition. If you happen to have a .22 conversion kit for your .223, you can use it for that as well.

Pistol Calibers: Plain old ball ammunition will work for all the major pistol calibers. Some .380 caliber ammo available is not up to the velocity required to penetrate the baffle, so be sure and use a good fully charged brand like Winchester, or American Eagle.

.223: Hole-punching ammo for a .223 requires a velocity that is faster than subsonic, but not by much. Subsonic ammo will not do for this purpose in this caliber. The velocity is too slow and what happens is that the jacket strips off and the lead goes through, leaving the jacket behind inside the unit. And it's a big headache to get it out.

Good .223 hole-punching ammo can be made from standard 55 grain. ball ammo. Don't use 62 grain SS109 type ammo for this, because the bullet with the steel penetrator won't work as well. Start by using pliers or a vise (or even a bullet puller) to work the bullet loose and pull it. Try not to mash the bullet or widen the case neck any more than necessary. Then pour out the powder onto a clean surface. Use a business card or something like that to line up the powder, Cheech and Chong style. Divide the powder carefully into fourths. Pour one fourth of the powder back into the case. Use pliers to tighten down the crimp a little bit again, and push the bullet back into place pointing forward.

A one-quarter charge provides a good velocity for hole punching. A one-half charge keeps the noise from being too excessive, but the faster velocity punches a perfect, .22 caliber hole leaving the baffle completely flat. That's great if you want flat baffles, but that's not what we want. We want the bullet to expand a little, causing the baffle to bulge out so we can make the indented baffle design. A one-quarter charge does that pretty well. It also keeps the noise down to where if rags are stuffed in the suppressor tube before firing, the noise level is less than a party popper so probably your neighbors won't complain. There is no need for pointing the gun upward or using filler materials inside the cartridge case. In fact, they seem to fire a little better pointed down.

.308: A similar procedure works for .308 as well. Just pull the bullet with pliers and reduce the charge to one fourth. Then, take a flat base bullet with a similar weight and push it in backwards, after crimping the case neck back down a little with pliers. In .30 caliber and larger, a flat base will cut a much cleaner hole than the pointy end. If you don't have flat base bullets handy, the boat tail base will work well enough.

.50 BMG: Start with a fired case. Use a 3/32" straight punch to drive out the primer. Use a utility knife to shave a little bevel around the edge of the primer hole. Then you can push a new primer into place and slowly beat it in with a hammer. (Wear safety glasses. And don't use a ball-peen.) Next, pull the bullet from a .308 cartridge and pour the whole .308 powder charge into the .50 case. Then, use pliers to work the case neck down tighter to where it will grip the bullet firmly. Put a

.50 Action Express pistol caliber slug into the case neck backwards and pound it into place. If you pack the end of the suppressor tube with rags really good and push it tight against the wood block, this is still quiet enough to do your hole-punching in your garage.

How to Build Military Grade Suppressors

Military Grade Suppressors for AR-15's

CHAPTER 3

The illustration on the following page is the basic internal design for most all the suppressors in this book. The baffle design consists first of a blast baffle, made from a heavy washer. This is made just to absorb the brunt of the muzzle blast. Then there are the freeze plug baffles, which are formed to an indented design that redirects gases across the center to disrupt the forward flow of the muzzle blast. The "indented" design is formed by first shooting through the baffle, causing it to bulge out. One side of the bulged hole is then pounded back down using a punch, and the hole is reamed with a drill and/or a round file.

Some commercial suppressors that use an indented baffle design have each of the indents alternating direction. My own tests showed that having them all face the same direction resulted in significantly better sound and flash suppression. The downsides were that in some calibers, having all the gases push one way can have greater effect on bullet flight. Another downside is that when rapid firing, more of the heat is directed to one side of the unit, causing it to expand and warp unevenly, sometimes leading to baffle hits at the front. The solution I came up with is to alternate the indents in pairs pushing the hot gas up and down.

A single hole is drilled in the edge of each baffle to relieve pressures on the side to which the gases are redirected. During development, I tried using baffles with no holes, and more holes, and results show that one hole seems to be the perfect number.

AR-15 Suppressor

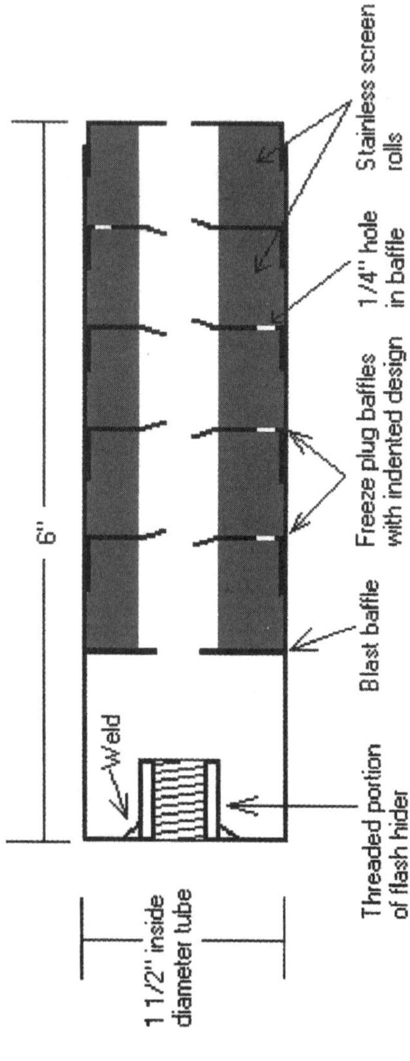

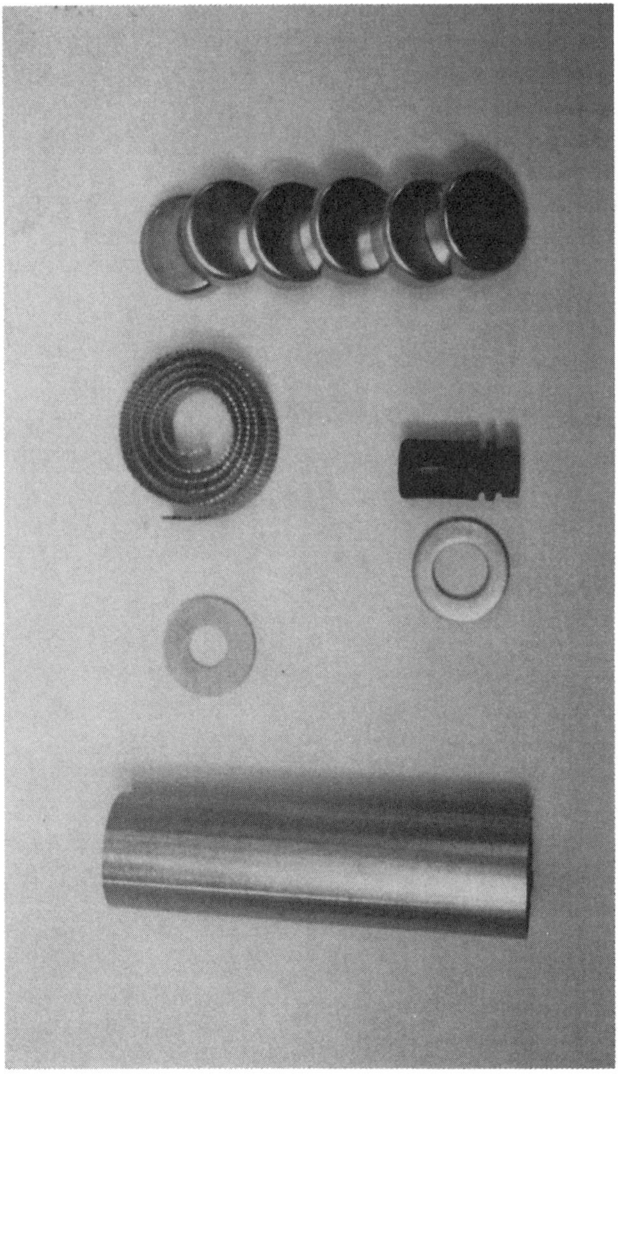

Fig. 3-1 Materials for construction; chain link fence pole, 1 1/2" inside diameter, 6" long; washer, 1 1/2" outside diameter, 1/2" to 5/8" inside diameter for blast baffle; washer, 1 1/2" outside diameter, 7/8" inside diameter for rear endplate; flash suppressor; strips of heavy stainless screen, about 3/4" wide, one for each freeze plug baffle; freeze plug baffles.

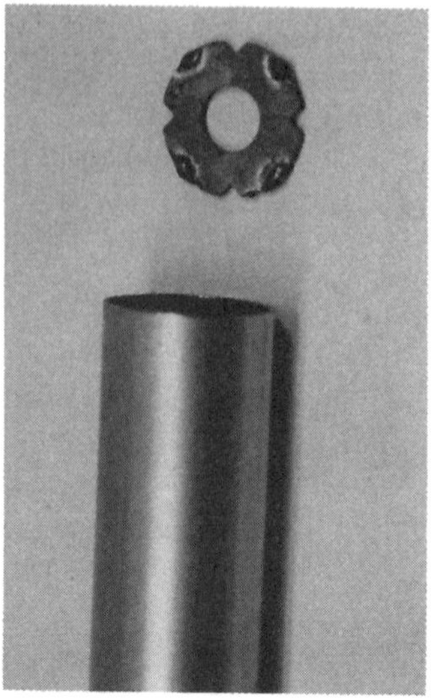

Step 1, (Fig. 3-2) Prepare the blast baffle. Grind four notches on the sides for pressure relief. If the washer fits loosely inside outer tube, use the welder to place four built-up lumps on the outside edge to hold it centered in the tube while being welded. Grind down the lumps so it fits snugly inside outer tube.

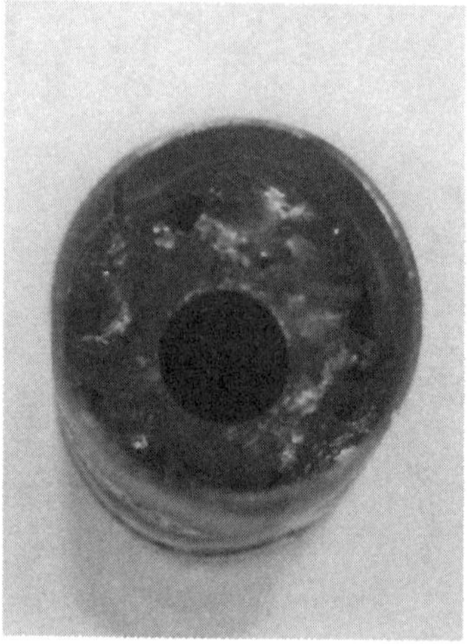

Step 2, (Fig. 3-3) Slide the blast baffle down into the rear end of outer tube about 1 1/2", and weld it in place firmly on four sides.

Step 3, (Fig. 3-4) Prepare the rear endplate. If needed, enlarge the hole in the rear endplate washer so that the flash suppressor fits snugly inside.

Step 4, (Figs. 3-5 and 6) Weld around the base of the flash suppressor. This is a critical weld, so it must be fused solidly. Use more than one layer of weld if necessary.

Step 5, (Fig. 3-7) Cut off the front half of the flash suppressor, leaving only the threaded portion. Then, use the welder to build up three lumps on the outside edge to hold the back endplate centered if needed. Grind the lumps down to fit inside the outer tube.

Step 6, (Fig. 3-8) Use a threading tap to clean up the threads. Be sure to use cutting oil. Do not attempt to thread a welded flash hider onto your weapon without doing this first, or you will probably damage the threads on the barrel.

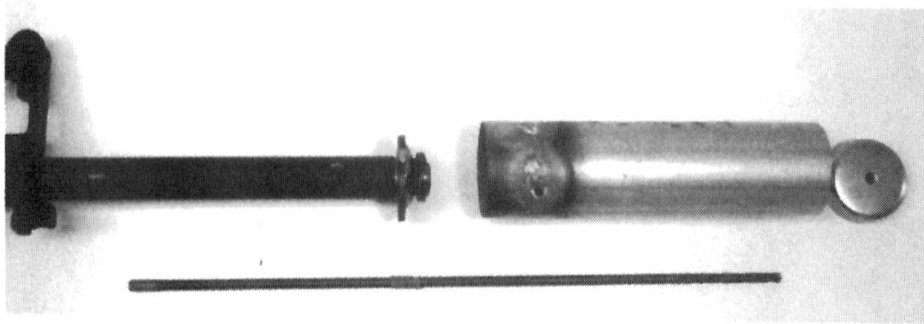

Fig.3-9

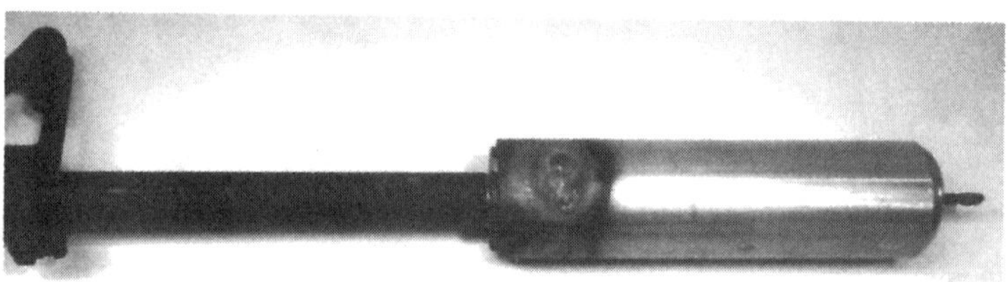

Fig.3-10

Step 7, method 1, (Figs. 3-9 and 10) Now it starts getting tricky. The next step is to place the outer tube onto the rear endplate and weld it in place, while keeping the outer tube aligned with the bore. There are a couple of ways to do this shown here. In the first method, begin by threading the rear endplate onto the muzzle. Then use a 12" long 3/16" drill bit for a centering guide rod. A 12" drill bit is used for this because it is much more reliably straight than anything else you could use. Use scotch tape built up at two points on the drill bit to make it fit snugly inside the bore. Place the drill bit in the bore, and then the outer tube onto the rear endplate. Then place the freeze plug shown into the end of the outer tube. The freeze plug has a 3/16" hole drilled in the very center, which the front end of the drill bit goes into as shown in Fig. 10. This setup will hold the outer tube more or less aligned while tack welding it onto the rear endplate. Wrap the barrel heavily in foil, and place it in a vise for tack welding. Note: This is not the preferred method. Scuff marks or welding spatter on the barrel can be difficult to avoid.

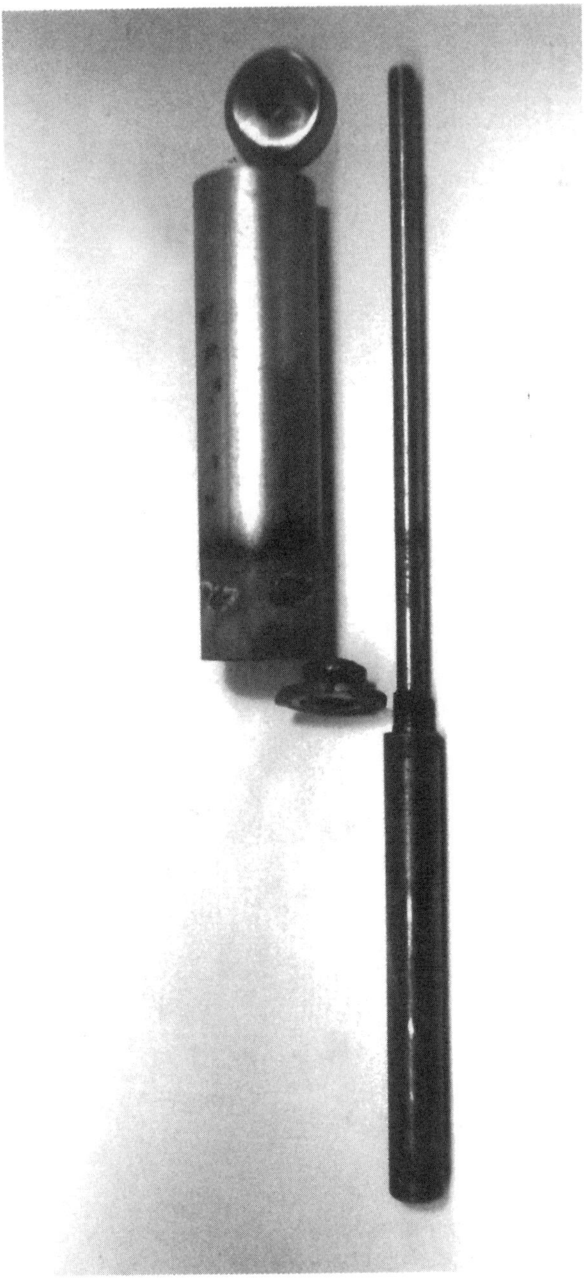

Step 7, method 2, This is the preferred method. It works in the same way, but won't risk damaging your barrel. Shown in (Fig. 3-11) above is a centering shaft, turned down from 3/4" roundbar to simulate the shape of a barrel with a 3/8" rod sticking out the front.

Step 7, Method 2 Cont'd. (Fig. 3-12) above shows the assembly put together in the same way as Fig. 3-10, and there is also a piece of conduit placed over the large end of the centering shaft to protect it from welding spatter.

Centering Shaft

Made from 3/4" steel round bar

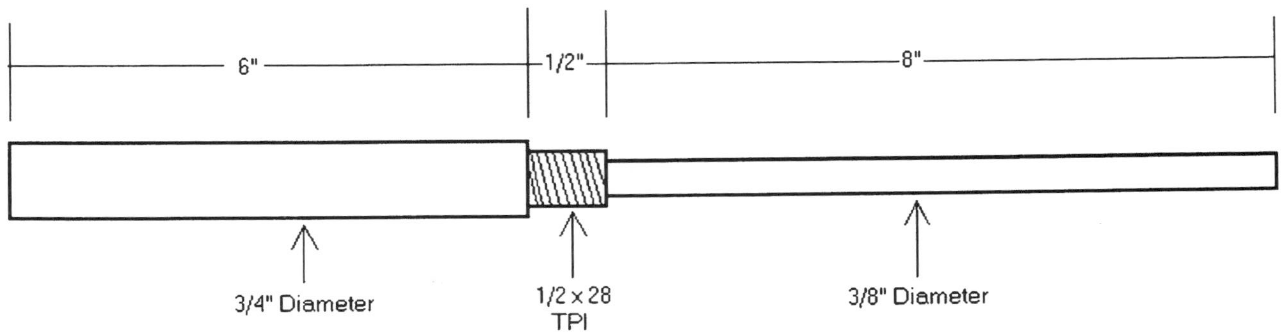

If you want to have your own centering shaft made by a machine shop, you can copy this drawing to take to them. Mine cost around $40.00 to have made.

Fig. 3-13

Fig. 3-14

Step 8, (Figs. 3-13 and 14) Fig. 13 shows what a corner-to-corner fit looks like. This allows you to fill in the corner space with weld metal, giving you a much stronger weld if done properly. After tack welding the endplate to the outer tube in three or four spots, remove the centering apparatus and complete the weld around the endplate as shown in Fig. 3-14. Use a bolt or something placed into the flash hider to cover the threads and protect from welding spatter. Grind the excess weld smooth.

Step 9, (Fig. 3-15) Place the unit on the barrel, and insert the 12" drill bit into the barrel to check alignment of the outer tube. If the end of the bit isn't perfectly centered in the end of the outer tube, make a mark on the side of the tube nearest the bit.

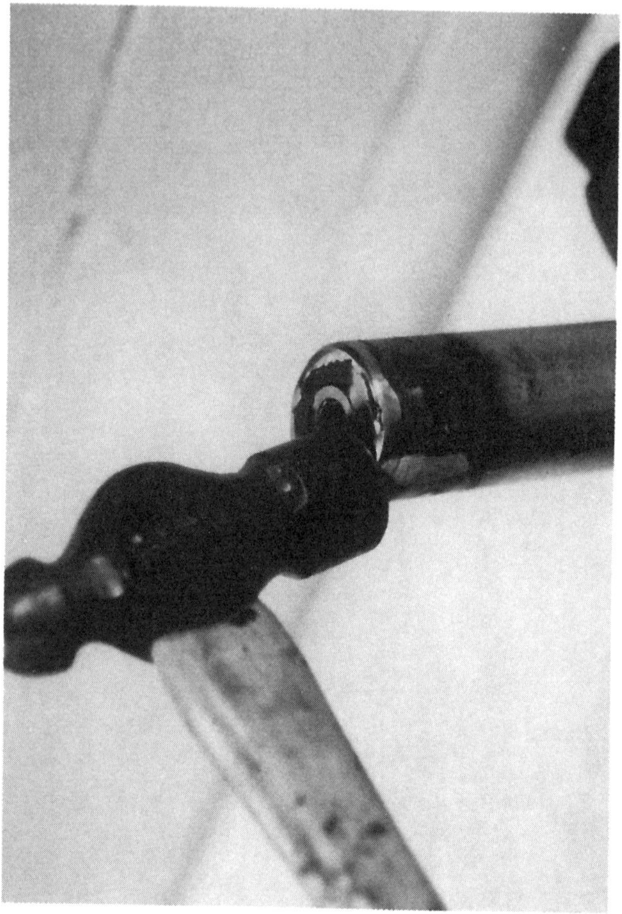

Step 10, (Fig. 3-16) Place the suppressor on a hard surface and hit the edge of the endplate with a hammer on the side where the mark is a few times. This will alter the alignment of the endplate on the outer tube. Place the unit back on the barrel and check alignment again. Repeat this process until it's straight and centered.

Step 11, (Fig. 3-17) Drill a 1/4" hole in the edge of the freeze plug baffles. When using the smaller 30mm tube and plugs, drill 3/16" holes. For huge suppressors like the .50 BMG unit, use 5/16" holes.

Step 12, (Fig. 3-18) Cut the strips of screen. When working with 1 1/2" baffles, I cut the strips 8 strands (3/4") wide, and 24" long. After cutting the stainless screen strips to the right length and width, peel out one strand of wire from each long side, and two from each end. This will keep it from unraveling inside the suppressor. Then roll it tightly around something like a 1/2" roundbar and place it inside the freeze plug.

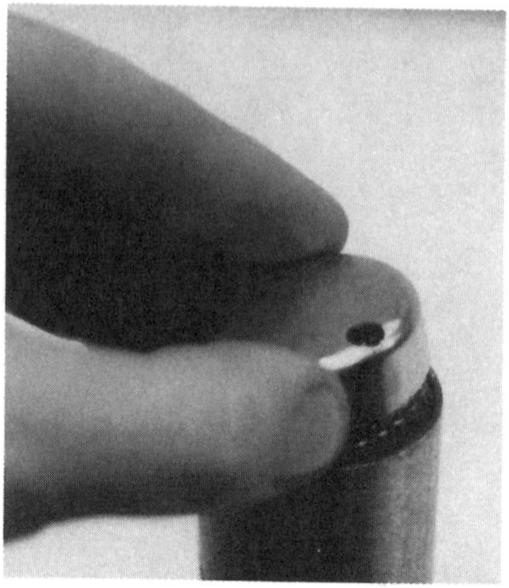

Step 13, (Fig. 3-19) Place the freeze plug baffle into the outer tube and force it to the bottom. Remember to place the holes so they follow the pattern in the design. Holes in the edge always go on the side where the indent in the previous baffle pushes the gases.

Step 14, (Fig. 3-20) Firmly tack weld each baffle on four sides. These tack welds must be firmly fused to both baffle and outer tube.

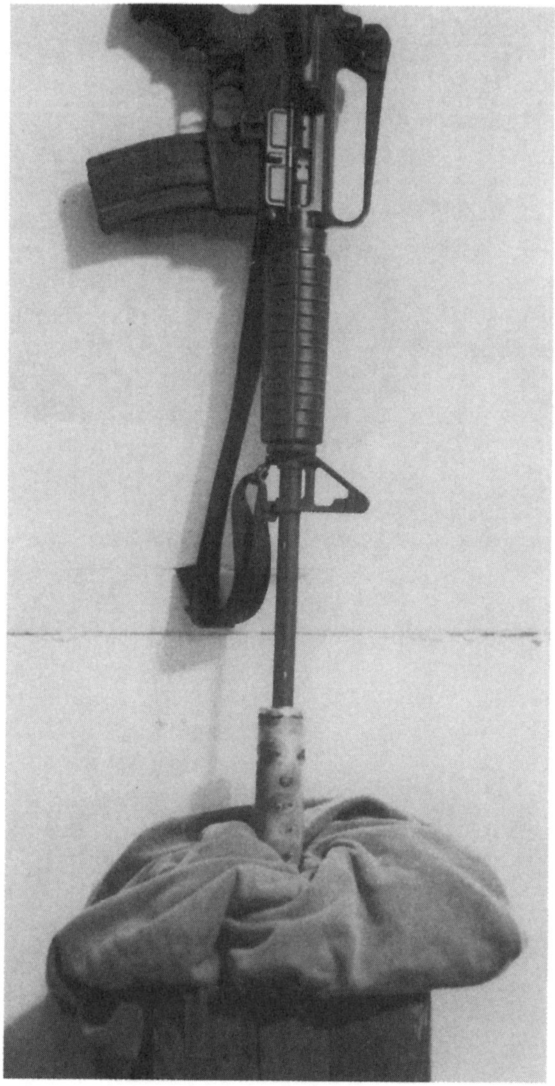

Step 15, (Fig. 3-21) Using hole-punching ammo, shoot the hole through the baffle. To do it relatively quietly without bothering the neighbors, stuff a rag into the end and shoot into a block of wood as shown, using special hole-punching ammo.

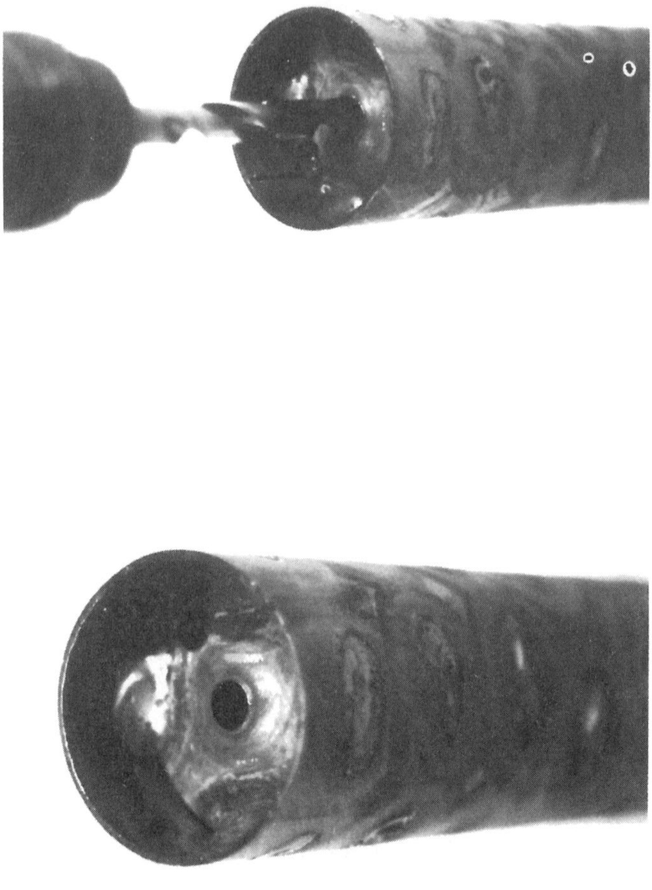

Step 16, (Figs. 3-22 &23) After using the hole-punching ammo, ream the hole with a 5/16" drill bit. Shake out any drill shavings.

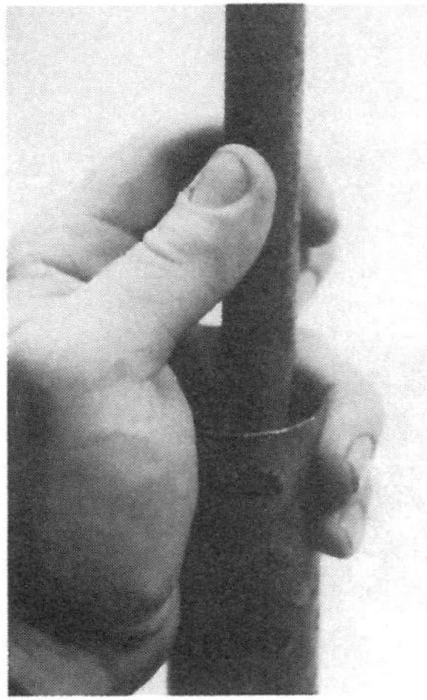

Fig. 3-25

Fig. 3-24

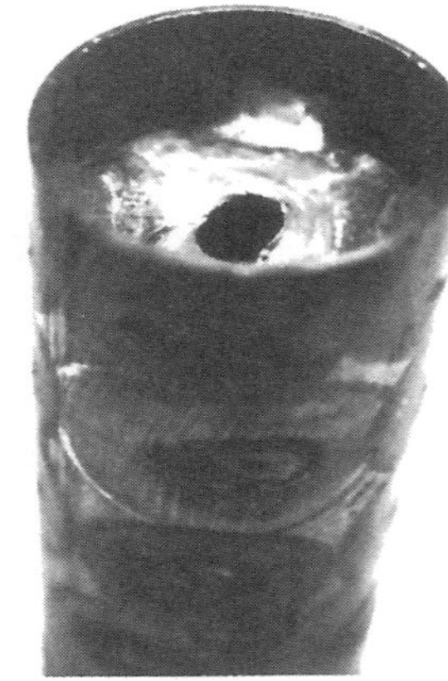

Fig. 3-26

Step 17, (Figs. 3-24, 25, and 26) Fig. 3-24 shows a simple punch made from 3/4" roundbar. Use this punch to pound the center hole flat on the same side as the drilled hole in the edge. Do not place the suppressor on a hard surface during this step. Hold the unit in your left hand, and hold the punch in place with the same hand as shown in Fig. 25. Rest the unit gently against your knee to steady it while you whack the punch with a hammer. Every couple of hits, re-ream the hole with the 5/16" drill. When done, the baffle should look like the indented design shown in Fig. 26.

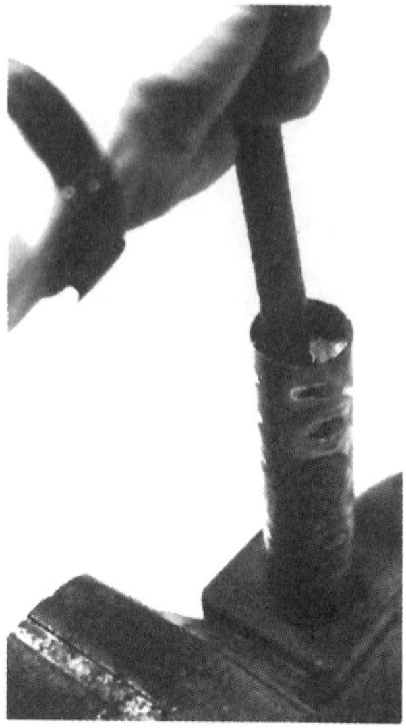

(Fig. 3-27) This is NOT the way to shape the baffle. Placing the suppressor on a hard surface will transfer the force of the hammering into the rear endplate, knocking it and the outer tube out of alignment. That's why it has to be held in the hand during hammering, to keep the force off the endplate.

Step 18, (Fig. 3-28) Since the side of the baffle that has been hammered down will be pushed closer to the center, you will need to use a round file or Dremel tool to enlarge that side so the bullet doesn't come too close.

Fig. 3-29

Fig. 3-30

Fig. 3-31

Step 19, (Figs. 3-29, 30, and 31) When you get to the end, place the last baffle partially into the tube, and place a sealing weld all the way around the outside. Then shoot the hole through, gradually pound it flat, and ream it with a 1/4" drill bit.

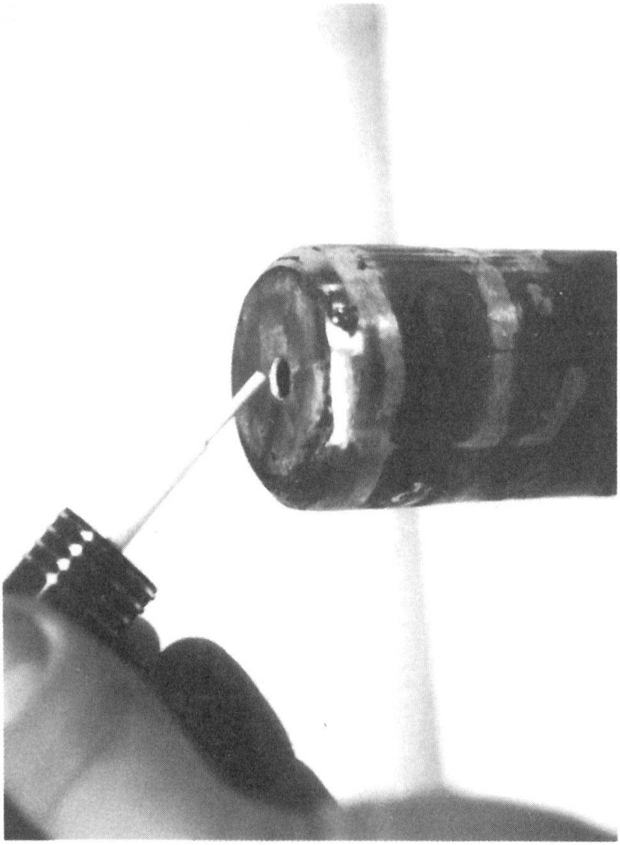

Step 20, (Fig. 3-32) To check the front hole for bullet hits, paint the inside edge with white nail polish. Then fire a subsonic or hole-punching round through, and check to see if the nail polish is rubbed away on one side of the hole, indicating a baffle hit. If so, use a round file to enlarge that spot until the problem is solved.

Step 21, (Fig. 3-33) Use Gun-Kote to coat the suppressor, inside and out. Use a container suitable for dipping, such as this shampoo bottle. Dip the unit halfway in, then let it drain. Take it outside and shake it out as much as possible, then hang it to dry. Repeat this process with the other half, then bake the unit at 300 degrees for 30 minutes or so. About three coats are needed, and the unit must be baked in between coats.

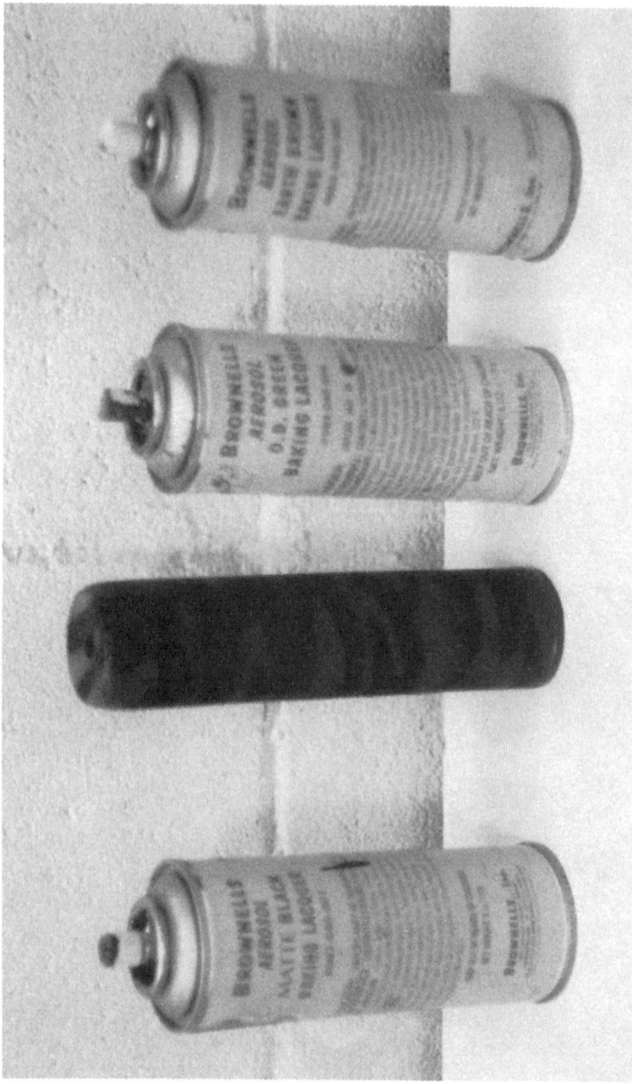

Step 22, (Fig. 3-34) If you like, you can use Brownell's Baking Lacquer to give the unit a camouflage pattern. An OD Green coat, possibly with shades of Earth Brown, and punctuated with streaks of Matte Black works well.

(Fig. 3-35) *Completed unit attached to an AR-15 rifle.*

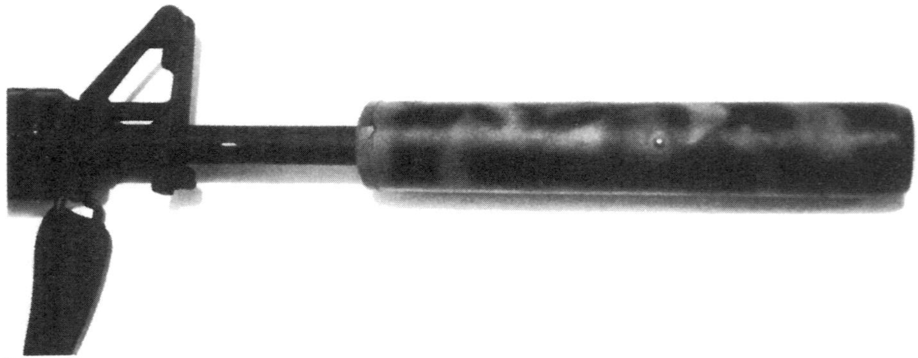

(Fig. 3-36)

The suppressor shown here is made from a flash hider that slips over a few inches of the barrel before it threads on, extending to the rear of the muzzle. This type of flash suppressor is available from Numrich Gun Parts Corporation. The concept is that by extending the expansion chamber back over the barrel a few inches, you create more space to absorb the muzzle blast before the first baffle, allowing a suppressor to be shorter from the muzzle. Suppressors that use a similar design concept are made by Reflex Suppressors. That company uses a different baffle design though, one which uses a cone protruding off the back of the baffle to direct more gases away from the center hole. We aren't currently able to do that with this construction method however. And though this longer unit extends 6" from the muzzle the same as the standard design, I was not able to notice a significant improvement in noise reduction. This type of design is most useful for use with a two-point mounting system, or when part of the barrel needs to be covered so it can be ported into the suppressor directly.

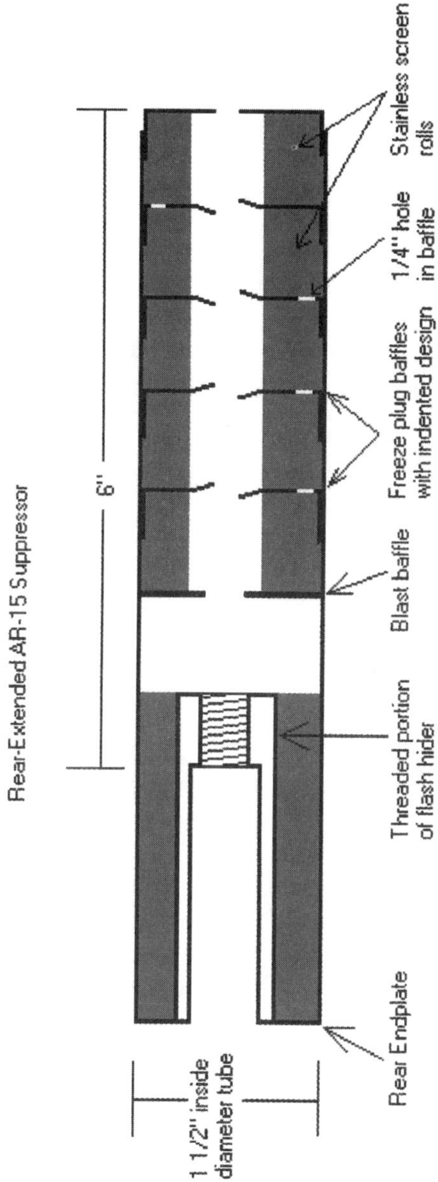
Rear-Extended AR-15 Suppressor
6''
Stainless screen rolls
1/4'' hole in baffle
Freeze plug baffles with indented design
Blast baffle
Threaded portion of flash hider
Rear Endplate
1 1/2'' inside diameter tube

How to Build Military Grade Suppressors

Upon Completion

CHAPTER 4

After completion of the unit, the next step is test firing, checking the function and alignment of the baffle holes, checking accuracy, and making any final modifications.

Because of the way the holes are made in the baffles, it is important that the suppressor be mounted properly, with good alignment, both during construction and use. An excessive number of spacers along with the lock washer on the barrel can alter the alignment of the suppressor during construction and/or use. If the alignment between the bore and suppressor is not absolutely perfect, the result can be bullets hitting the baffles. While this will not cause it to explode or anything like that, it's still a very undesirable thing. A single lock washer on your AR-15 will not hurt the alignment. If your AR-15 requires a lot of spacers along with the lock washer for the A2 flash hider to line up properly, then you might consider solving that problem by switching to an A1 or Vortex flash hider to eliminate the need for extra spacers.

If the construction procedure has been followed carefully, then the alignment of the baffle holes to the bore should be OK already. If at any time you think it might have been knocked out of alignment, then do the test with the white nail polish inside the front hole and a round of subsonic or hole-punching ammo.

When you first take it to the range for test firing, the first step will be to test fire it with full power ammo and see how it holds up under stress. Start by firing a few rounds through and then unloading the weapon and checking the condition of the unit. Examine the front hole for any signs of bullet hits. If you see any, then use a Dremel or round file to enlarge the part of the hole being hit. Bullet hits on the front or any other baffles have to be fixed right away before further firing. These bullet hits put a lot of stress on the weak point of the unit, which is the weld between the back endplate and outer tube. They can knock the entire unit further out of alignment very quickly, especially in pistol calibers. One time I was testing a .380 caliber suppressor that had this problem, and I decided to fire several more rounds until I could see the baffle hits more clearly. Suddenly the entire unit just dropped off, leaving only the back endplate attached to the gun. It was a little embarrassing to say the least. This is why I am now careful about not grinding the weld too thin, and also why I now fit and weld the back endplate corner-to-corner to the outer tube. Baffle hits anywhere inside the unit also leave bullet shavings inside the suppressor. After they start to gather up, they can blow back into the gun, causing it to jam up in ways you'd rather not know about. The reason this happens is that there is still pressure inside the unit when the breech of the gun is open, and so some of that pressure blows back through the barrel carrying debris, moisture, and every other type of crud from inside the suppressor with it.

When you're certain of your alignment, then you will have to rapid-fire the unit to see how it holds up to the heat. This is an important step with rifle calibers. With the weapon on semi-auto, fire two or three clips through rapidly. You can do more if you want, or do it full auto, but you'll fry the coating and have to clean it and coat it over again. After the unit cools a bit, check the front hole again for signs of baffle hits. The extreme heat can cause the unit to warp to one side or the other, resulting in hits on the front baffle. If you see signs of any, enlarge the affected part of the hole just to the point where it stops hitting.

After this phase is complete, the next step is to check accuracy. First, get the rifle zeroed at a relatively short range, like 25 yards, and determine the normal group size with no suppressor. Then, attach the suppressor and fire another group. The location of the group might be a little lower or to the side, but the size of the group should be the same. If it's not, check further for baffle hits. Typically, one of these suppressors on an AR-15 with a heavy barrel will cause the group to be about 4 inches off at 100 yards.

If you fire your rifle rapidly, the extreme heat can also bake off the galvanizing on the metal parts. This will cause an unpleasant stink, and contribute to fouling of the gun. One method of prevention is to remove all the galvanizing by burning it off with a torch, or bead blasting the parts before assembly. That's a lot of extra work though, and I generally don't bother with it anymore because it's not a huge problem anyhow.

To increase the suppression effect, the unit can be fired "wet." What this entails is simply inserting 5 to 10 cc's of water into the suppressor. With this suppressor design, the water can be inserted from the front, and then with the suppressor tilted back, the water will run down into the unit through the gaps between the welds on the baffles. An easier way, when feasible, is to just dip the suppressor in water and drain it out again before firing. The water helps by cooling the muzzle blast gases. Then as it turns to steam, the steam helps extinguish any unburned powder, reducing the blast, until the water is all evaporated and gone. If you plan to fire it wet very much, you definitely want an internal coating of Gun Kote to protect from corrosion. Also remember that water, like anything else floating around in the suppressor, can blow back into the gun, so don't forget to take precautions against corrosion when you're done shooting. Another substance people use to increase suppression is lithium grease. Placing some around the inside of the front and rear end of the unit will have an effect similar to water.

After a great deal of shooting, your suppressor can become clogged with fouling and lose effectiveness. If your suppressor is internally coated with Gun Kote, you can easily clean it by immersing it in soapy water, then plugging both ends and shaking it vigorously. Then drain it, rinse it out with hot water, and dry it out by baking it at between 200 and 300 degrees until it is completely dry. You can also use denatured alcohol as a cleaner. Just plug one end with your finger, pour some into the unit, then plug both ends and shake it, then drain it. The alcohol will then evaporate fairly quickly. Another cleaner you can use is carburetor cleaner.

There is a major consideration to make if you plan to use the suppressor in a genuine battle situation. A suppressor causes the gun to retain a lot of the fouling that would normally go out the muzzle, and the gun will require cleaning a lot sooner than normal. You can still expect to get several hundred rounds out before the gun starts to clog up, but a person going into an extended battle is probably better off without one. Suppressors are better suited for snipers and short-term combat situations, like a raid. They certainly do make it easier to do your work though, because you don't need hearing protection, thus you can still communicate more easily, and you don't have to deal with that abominable roar that an AR-15 or similar rifle can produce.

One last safety tip. While it might be tempting to stand downrange and listen while your buddy zings bullets past you, NEVER do this. Not even if your accuracy is still perfect. I have seen tracers deflected by as much as about 30 degrees by baffle hits, and there are many reasons why this could happen unexpectedly. These include the suppressor shooting loose from the barrel, bad welds cracking, and heat warpage, among other reasons. A much smaller deflection than that, and you're nailed.

Shooting Results for AR-15

CHAPTER 5

A suppressed AR-15 is a joy to shoot. The inherent versatility of the weapon, its quality and accuracy, along with the variety of high performance and subsonic ammo available make a suppressed AR-15 one of the most all around lethal personal arms you can buy. Another great thing about this rifle is that you can get a lot of accessories that actually work well with the rifle. You can get scope mounts that actually hold their zero, good night sights, weapon lights and lasers that integrate well with the weapon. That's because the rifle is a real military design that is made to accept these things, and the same things can't be said for aftermarket accessories for a lot of other weapons.

The only real complaint about the suppressed .223 is that because of the velocity of the bullet, the supersonic crack is rather loud. Velocity has a rather large impact on the level of supersonic crack. For that reason, a suppressed .308 is actually less noisy than a .223 with standard ammo. The amount of noise generated depends on the distance traveled by the bullet. Shooting a target right up close is a lot less noisy but still makes a pretty good "thump," much of it coming from the bullet impact.

There is one type of supersonic ammo designed to reduce this problem, which is called "Blue Tip Reduced Charge." This is a military ammunition load using a 62 grain SS109 type bullet, loaded down to around 2000 fps. The problem with this ammo is that although the noise is still substantial, the velocity is reduced below the point where the bullet would fragment on impact with the target. This means that you're trading a huge amount of effectiveness in exchange for not a great deal of noise reduction.

With standard supersonic ammo, accuracy with the suppressor attached remains equal to accuracy without, the only difference being that, with an AR-15 anyhow, the group is usually displaced by about 4 inches or so. The displacement can be in any direction, up, down, or to the side. The change in location is caused by the change in barrel harmonics brought about by the weight of the suppressor. Exactly how the barrel is vibrating when the bullet leaves the bore helps determine the point of impact. Heavier barrels are therefore less prone to this problem and will require less of a change in zero. Another factor is how the internal turbulence pushes on the bullet. For that reason, whenever possible I try to make the indented baffles push gases straight down, though this isn't always feasible, or really all that important so long as you practice enough to know how your weapon will place shots with and without the unit.

Subsonic ammo is, of course, the way to go for stealth. There are two good brands currently available in .223, they are Engel Ballistics and TTI Armory. Neither of these will cycle the action of a .223. There is also now a new brand of .223 subsonic that will cycle an AR-15, and stabilize the bullet if the barrel has a 1x7 twist. It's called the Advanced Subsonic Round and is available from Extreme Shock Ammunition for a small fortune.

When I test fired the .223 suppressor, I found that the accuracy actually improved a little bit with the suppressor on the rifle. The reason for this is because the weight of the unit reduces the barrel vibration that so often is the Achilles' heel for good accuracy. The center of the group on this particular unit was centered about 4 inches to the left of the group fired with no suppressor. When firing with Engel Ballistics subsonic ammo, the groups were centered approximately 24 inches lower than standard ammo at 100 yards. Standard ammo groups in my rifle ran about 1 1/2" to 2". Engel subsonic ammo groups were about the same size.

For torture testing, I tried rapidly dumping four magazines of full power .223 through one of these suppressors, using a full auto M-16 with an 11 1/2" barrel. When I was done, the coating was burned completely off, but structurally the unit held up fine. After cleaning and re-coating, it was like new again. More or less.

How to Build Military Grade Suppressors

More Designs .22 thru .50 BMG

CHAPTER 6

AR-15 in .300 Whisper Caliber

The .300 Whisper is a .223 type case that is necked up to a .30 caliber. It can be loaded with lighter bullets for supersonic loads that approximate the 30-30, and also loaded with heavier bullets from 220 to 250 grains for subsonic loads. Ammo in this caliber is made only by Cor-Bon.

This caliber currently has the distinction of being the only American subsonic rifle caliber that will function in a gas-operated weapon. Like most exotic calibers, it has its ups and downs. On the upside, it gives you the most punch for the noise level you can ask for in a rifle caliber. It doesn't have any breech blast problem to increase noise, and also doesn't require manual cycling like other subsonic rifle caliber ammo.

On the other hand, ammo costs a fortune, even if you load your own, and is not what you'd call a commonly available caliber. The only upper receiver made to cycle factory .300 Whisper ammo is made and sold by J.D. Jones of SSK Industries. Special dies have to be purchased from him to make your own brass from .221 Fireball brass, and he only sells them to people who tell him they have one of his AR uppers. Olympic Arms makes a 1:8 twist .300 Fireball barrel that works with Whisper ammo, but only special handloads will cycle the action of their uppers.

While the caliber is advertised as being sniper-accurate to

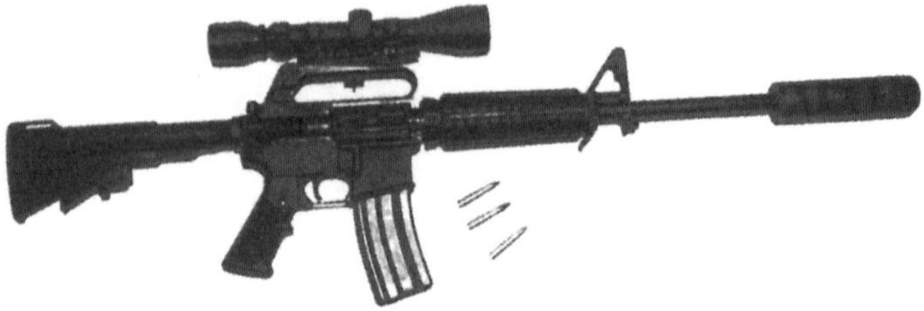

(Fig. 6-1) AR-15 in .300 Whisper caliber

300 yards, I think they are probably referring to accurate bolt-action guns made for it. With an AR-15, I got 3" to 4" groups at 100 yards with Cor-Bon 220 grain subsonic ammo, and the groups spread out rapidly from there. That makes 100 yards about the maximum range I'd want to reliably pick off a target with an AR-15. The supersonic 150 grain ammo gave me 2" groups.

Probably the best military type use of this weapon is for an assault where some measure of stealth is required, at least until you're detected, at which time you can switch to more effective supersonic ammo. This gun is a lot of fun for recreational shooting as well, and does a pretty good job on critters.

AR-15 in Pistol Caliber

The AR-15 shown on the next page uses a pistol caliber upper receiver and a barrel that has been cut down to 6". The suppressor is an integral type and is made from the 14" steel handguard tube.

Steel handguard tubes are very hard to find these days, as everyone is switching to aluminum and other lighter materials. Olympic Arms occasionally has them though, and they can even make you one by special request. About the only other option I know of is to try and steal one off of an existing rifle and replace it with an aluminum one.

(Fig. 6-2) AR-15 in pistol caliber with integral suppressor built into handguard.

Olympic Arms makes magazines for use with their pistol caliber upper receivers that are made from various types of surplus hi-cap magazines. They have steel enlargements welded to the top of the mag to fill the space in a standard AR magazine well. That means that the mags won't fit into regular pistol caliber mag pouches unfortunately, but will require .223 mag pouches. Olympic used to make conversions that allowed you to use Glock magazines, but unfortunately they don't any longer.

(Fig. 6-3)

Fig. 6-3 shows the gun with the handguard / suppressor removed. The barrel has been cut to 6" and recrowned. The barrel nut portion of the handguard is full of holes that have to be plugged, and this is done with epoxy gooped into the holes with your finger from the inside, while carefully keeping the epoxy off of the barrel nut threads. The baffles are welded in starting an inch or so in front of the muzzle. 1 3/4" is generally the inside diameter of the handguard tube. No blast baffle is needed for pistol calibers.

Accuracy and function of this weapon are good, and the muzzle noise suppression is excellent. There is some of the increased noise from the breech blast though, as this upper receiver is recoil-operated. You can see the difference for yourself by holding a rag over the ejection port while firing. Because the barrel is so short the problem isn't bad, but it's enough to make the .300 Whisper preferable where stealth is critical.

Springfield M1A

The .308 caliber seems to be very amenable to being suppressed. If you fire this suppressed M1A rifle in the air with nothing around to reflect supersonic crack back at you, you will hear a sound roughly equivalent to a cough. With rifles like the M1A or FN-FAL, you can shut off the gas system so it works like a bolt action, eliminating action noise and making it even quieter.

This suppressor was tested using standard ammo and also with TTI Armory 168 grain subsonic ammo. The first tests were

(Fig. 6-4) Springfield M1A with 18" barrel and 8" suppressor

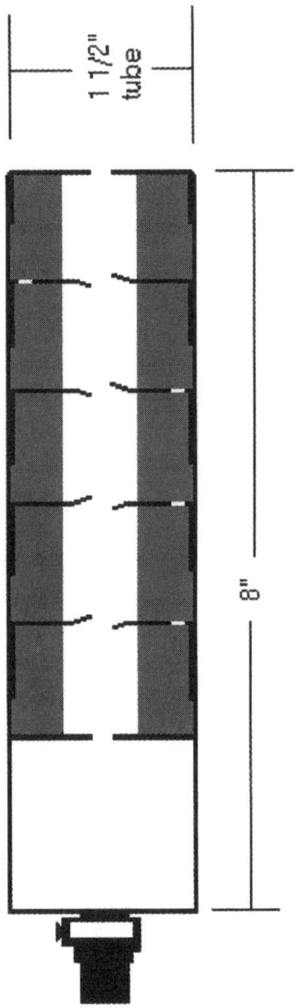

Springfield M1A Suppressor

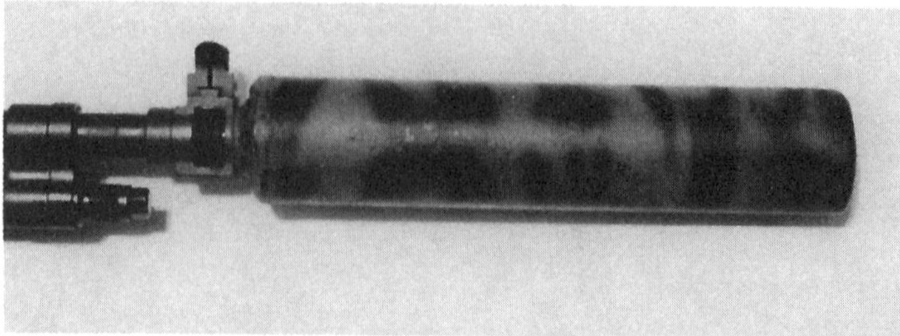

(Fig. 6-5) Close-up of M1A suppressor. Note that front sight is high enough to still be used.

done with an M1A that had a full-length 22" medium weight barrel on it. The groups appeared about 8" lower than without the suppressor. Later, I used the suppressor on an M1A that had the 18" standard (light) weight barrel shown in the photos. With this barrel, at 100 yards the groups appeared about 14" lower and a few inches to the left from where they were without the suppressor. Suspecting there was something wrong with the unit, I cut it apart and rebuilt it entirely, and got the same identical results. Since the Springfield flash hider slides onto three grooves, I also tried turning it 120 degrees each way. The point of impact changed very little. This means the cause can only be the weight of the suppressor affecting barrel harmonics. Evidently with this type of gun you'll want to go with a heavier barrel than the standard weight. Otherwise you might have to either make it a dedicated suppressed weapon, or have interchangeable scope mounts with different zero points (that hopefully hold their zero when you change them.)

The TTI Armory subsonic ammo produced good groups about 18" lower than standard ammo at 100 yards. This ammo also had about the same difference in group location when the suppressor was added or removed.

At the time of this writing, Engel Ballistics was working on a .308 subsonic round that would stabilize in a regular 1:12 twist .308 barrel, and also cycle the action. I sure hope they can pull that off, that would be some really nice ammo to play with.

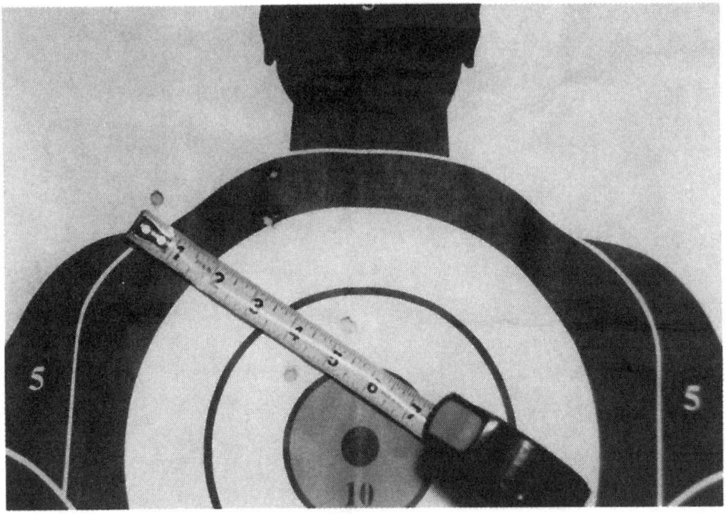

(Fig. 6-6) This M.O.A. group was shot at 500 yards through home-built suppressor on the M1A.

Makarov .380

For stealthy and effective pistol shooting in something bigger than a .22, the Makarov in .380 is currently the best thing going. I've worked with both Russian and Bulgarian pistols and they both performed outstandingly. If you can, try to find a Russian pistol with the double stack 12 round capacity. The magazines for that version now come with a stop inside that makes it a 10 rounder, but that's easily fixed.

This pistol is every bit as reliable with a suppressor attached because the barrel is fixed on the frame. You will need to purchase an aftermarket threaded barrel from Federal Arms and install it.

Fig. 6-7 shows two suppressors, one made from 1 1/2" tube and one made with 30mm tube, and both are 6" long. The performance of the smaller unit is not bad, but for really good performance go with the larger size.

White nail polish is used on the front and rear sights for visibility. Aiming accurately is still possible even though the

(Fig.6-7) Makarov with 30mm wide suppressor attached, and 11/2" suppressor below.

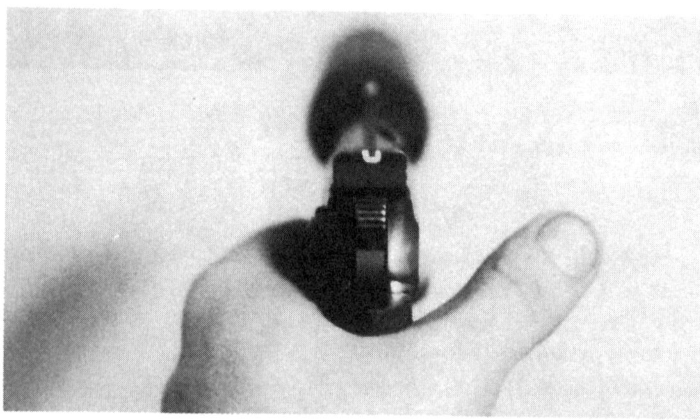

(Fig. 6-8) Sight picture using painted sights

suppressor blocks the line of sight. Just aim with both eyes open, looking at the sights with your right eye and focusing on the target with your left eye. This takes some practice, so plan on spending some time at the range working on this technique. When coating the suppressor with Baking Lacquer in camouflage colors, be sure to keep the back end of the suppressor black so the whitened sights will show up more clearly.

This pistol is a lot of fun to shoot suppressed. Accuracy is very good, and with careful shooting the gun will print nice, tight groups. The Federal Arms barrel improves accuracy over the original barrel, as the chamber dimensions are tighter. When punching holes in the baffles with this gun, remember to use heavy leather gloves to hold the slide closed to avoid damage from excessive pressure and impact of the slide blowing back.

Ruger 10/22 Slip-on Muzzle Suppressor

This suppressor is made from a regular slip-on flash hider with a set screw. The tube is 30mm by 6" long. The set screw has been replaced with a longer screw, and a wing nut has been threaded on and tack welded in place at the bottom end. This

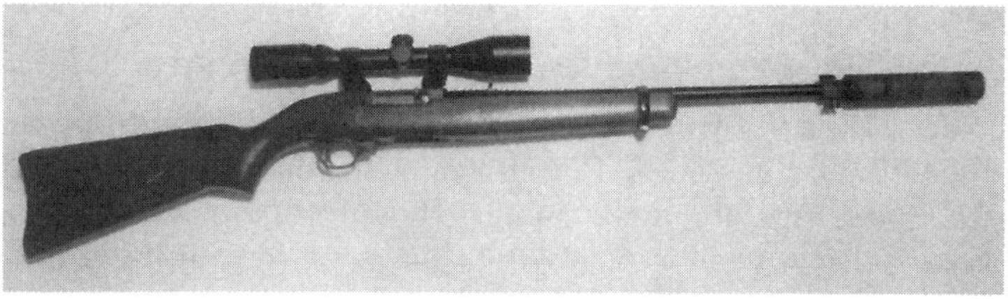

(Fig. 6-9) Ruger 10/22 with slip-on muzzle suppressor.

(Fig. 6-10) Closer view.

allows you to tighten it or remove it by hand without need for tools. The .22 long rifle cartridge is light enough that the set screw by itself will hold the suppressor on just fine, with no need to drill an indent into the barrel.

This gun was tested first with CCI Standard Velocity ammo. This ammo was a little too fast in the 16" barrel and produced a ballistic crack. The next slowest ammo was Lapua Scoremax. This is a barely subsonic ammo that uses a heavier 48 grain bullet. At 100 yards, it grouped only about 3" lower than the standard velocity, meaning it's running at a pretty optimum velocity. With a standard full length barrel, this would be my number one choice for head shots on large game. It would be a far better choice than other types of subsonic, which are both slower in velocity, and lighter in weight. The only downside is that this ammo is a little on the spendy side, to the tune of about $65.00 for a 500 round brick. However, the points of impact between this ammo and standard velocity are close enough that you can use inexpensive standard velocity for target practice, and the Lapua Scoremax for field work.

Ruger 10/22 Integral Suppressor

Yes, I know this has been done before. This concept for an integrally suppressed 10/22 was created by Mark White of Sound Tech Industries, and many kudos to him for doing so. However, I've created a version of my own that's worth putting in this book.

(Fig. 11) Integrally suppressed 10/22

This version is built around a rear-extended AR-15 flash hider. The 10/22 barrel has been cut to 6" overall length, and turned down to a design that creates an integral two-point mounting system for the flash hider that the suppressor is built on. One benefit of this system is stability. The suppressor will not wobble at all, and barrel vibrations are significantly reduced. Another is that the barrel and the flash hider can be ported to vent the gases into the suppressor if desired. This is useful if you plan to use Aguila SSS ammo with the gun, as this ammo requires two 1/8" ports be drilled into the barrel 2" back from the muzzle for the 60 grain bullet to stabilize.

On the subject of Aguila SSS ammo, I'll make no bones about it, I'm not a fan of this stuff. First, it has a lower-than-optimum velocity to begin with, out of a 16" barrel even. Second, the short case tends to clear the chamber before chamber pressures have dropped, substantially increasing chamber noise. The barrel has to be 6" or less to alleviate this problem, reducing velocity further. Then the barrel has to be ported to reduce the muzzle turbulence in order for the 60 grain bullet to stabilize, by which time you have a trajectory not unlike that of a golf ball. That means that at anything past about 50 yards you have to be very precise about your range finding and bullet drop compensating, and that means a lot of range time shooting at every different yardage before you can put it to use. I should mention, in fairness, that the people from Sound Tech report having great success using this ammo to dispose of deer out to 110 yards with their rifles that are made for it. Still, if Aguila could bump up the velocity to a little over 1000 fps, they would have a much more useful round, even if it meant reducing the weight of the bullet a little.

One time I asked a trapper to test this ammo with his standard 10/22 on some newts, which are the largest rodent in North America. He reported that at any range over a couple of feet, the bullet was completely ineffective and did not provide an instant kill because of the instability in an unmodified barrel. If he held the muzzle right up close, he got an instant kill.

Another type of ammo I hate is Remington .22 subsonics. This cartridge is low powered and ineffective, yet at the same time it's one of the noisiest types of subsonic ammo. How they achieved this I don't know, but this round generates a louder pop from the ejection port than other types of subsonic. Also, every now and then a round goes supersonic, even from the short barrel, decimating any stealth.

Luckily, for this rifle, the perfect ammunition is either CCI or Remington 40 grain Standard Velocity ammo. In a 6" or 7" unported barrel, standard velocity ammo runs at an optimum subsonic velocity. This ammo did create a ballistic crack with the 16" suppressed 10/22, but not when used in the short barrel. Standard velocity is also accurate and cheap, has a much flatter trajectory than other types of subsonic ammo, and has the effectiveness you need, not to mention being far more reliable in a semi-auto. Once I shot up an abandoned car with both Remington subsonics and CCI standard velocity. The Remingtons bounced off the skin and the Standard Velocity punched right through.

Since the effective range of this gun is about 100 yards, a good range to "battle zero" it at is 75 yards. If zeroed at that range, standard velocity ammo will group about another 4" lower at a full 100 yards.

The drawing on the next page shows the design of the barrel made for this suppressed 10/22. You can have it made in a machine shop by just chopping a standard Ruger barrel down and taking it to them with a copy of this design drawing. Leave a few extra inches of length for them to work with before they cut it down to the final 6". If you want to make it a little longer, about 7" overall is the longest you can go with a standard Ruger barrel, before it tapers too narrow.

If you tell the machinist making the barrel that it's going onto a pistol-configuration gun, there are no paperwork issues related to making a short-barreled rifle. After it's made you can coat the bare, machined parts on the outside only with Gun-Kote using a cotton applicator. Do not coat the bore.

This barrel is designed for an integral suppressor made from a rear-extended AR-15 flash hider from Numrich Gun Parts Corporation. The flash hider threads on at the front, and the back end of the flash hider seats up against the 45 degree portion of the barrel, creating a rear seal and an integral two-point mount for extra stability.

The threaded portion of the barrel is extra long, 3/4" so that when the suppressor is fully threaded on and seated at the rear, there is a little gap remaining between the inside threads of the flash hider, and the non-threaded part of the barrel. This is so that the threads will pull the flash hider firmly into the 45 degree part of the barrel for solid two-point mounting. This also gives the barrel a little "stretch," enhancing accuracy.

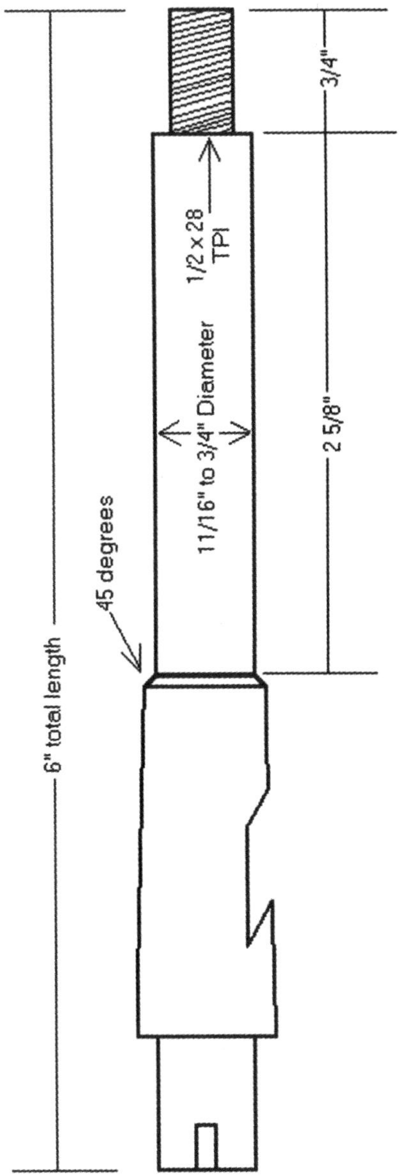

10/22 Barrel
Lathe-turned from cut-down factory original barrel

This barrel can be optimized for use with Aguila SSS ammo if you want. The 6" overall length shown in the drawing is the right size to allow chamber pressures to drop before the case clears the chamber, keeping the breech noise down. The barrel can also be ported for stabilizing the 60 grain bullet. To do this, just drill a 1/8" hole through both sides two inches from the muzzle. This will reduce the muzzle turbulence just enough to allow the bullet to stabilize. If you port the barrel, you will have to also put matching ports into the slipover flash hider prior to building the suppressor around it, in order for the gases to vent into the suppressor. Since the ports in the flash hider will move as the threads get worn in, elongate the holes in the flash hider laterally to allow for that. Another way to stabilize the 60 grain bullet would be to use an aftermarket barrel that uses a rifling twist tighter than the standard 1x16 twist used by Ruger. Aguila is now making their own barrels with a 1x9.5 twist, but they're spendy.

If you decide to port your barrel, but want to be able to close the ports for using Standard Velocity ammo, you can thread the ports and plug them with removable set screws. Of course, this porting and threading will not do much good for accuracy. Unless you're intent on using the Aguila SSS ammo, you're best off leaving it unported and using Standard Velocity ammo, which runs at the perfect barely subsonic velocity from this barrel, or the Lapua Scoremax if you want a heavier bullet.

The design of this suppressor is similar to the Rear-Extended AR-15 suppressor shown at the end of Chapter 4. This 10/22 suppressor was made from 1 1/4" thinwall conduit. The actual inside diameter of this tube was more like 1 3/8". The length extends 7" forward of the muzzle. Sound suppression is outstanding. It is very significantly quieter than the slip-on muzzle suppressor for the full-length rifle, and very much quieter than a suppressed AR-15 with subsonic ammo. This gun is 100 yard head shot accurate with standard velocity ammo.

For more tips on working over your 10/22 for accuracy, reliability with subsonic ammo and so forth, pick up the book "Ultimate Ruger 10/22" by Mark White, available from Delta Press.

Also, soon Aguila will be coming out with a new subsonic .22 magnum round called the Goliath. This round will have a 75 grain bullet. Presumably they will be making barrels with a tighter twist for use with this round as well. It should be very interesting to work with.

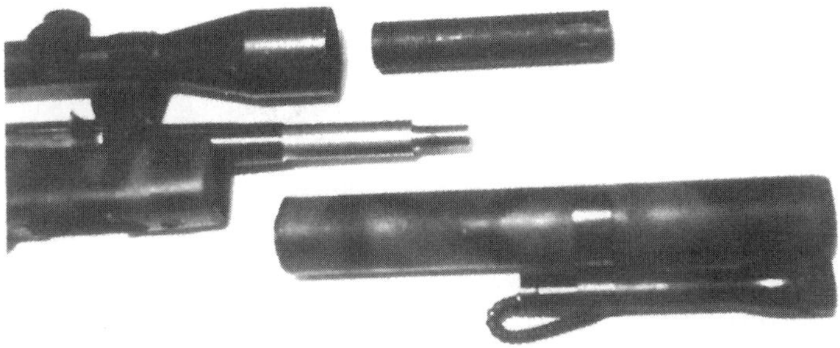

(Fig. 6-12) Close-up of 10/22 barrel with suppressor detached. Unmodified rear-extended flash hider shown above.

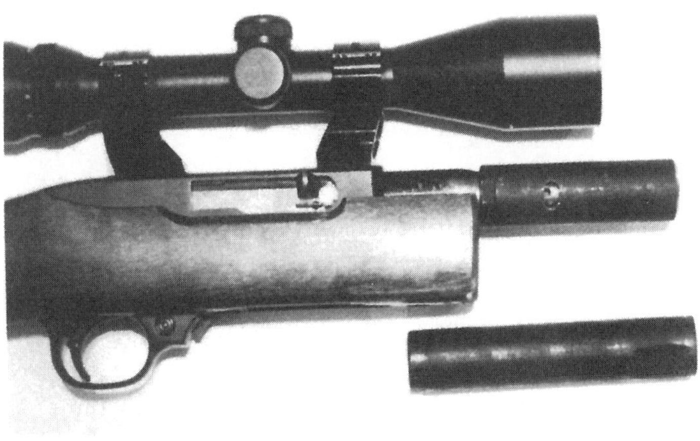

(Fig. 6-13) 10/22 barrel with rear-extended flash hider attached. Flash hider is cut off forward of the threads, and a full size flash hider is shown below. Note the ports in the barrel, with matching holes in the flash hider to vent gases into the suppressor once it is built around the flash hider.

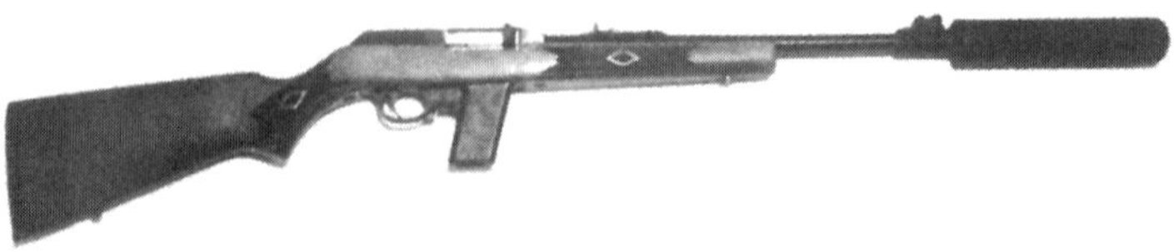

(Fig. 6-14) Marlin Camp 9 with suppressor built from slip-on flash hider

Marlin Camp 9

The suppressor shown on this rifle is built on a Choate slip-on flash suppressor. A shallow hole had to be drilled in the barrel where one of the set screws goes in order to stop the suppressor from being blown off the barrel. The standard size set screw didn't like this very well, as the battering mashed and deformed the threads, making it impossible to get out without drilling it out. The solution to this would be to bore out the hole in the flash hider to slightly over 1/4", and weld a 1/4" nut on the outside of the hole for a larger, preferably hardened steel set screw. After welding, the nut would probably need the threads cleaned up with a tap. Then you would have to drill a 1/4" indent in the barrel to accommodate it.

The other problem with this system is the barrel length, which creates a lot of chamber pressure and a loud breech blast, which unlike supersonic crack, will point the opposition right to your location. The breech blast was a substantial problem even with the lightest 124 grain subsonic load I had.

The bottom line on this system is that it can be done, and will work fine for a plinker or bunny gun. But for a suppressed weapon system, there are far better choices to work with.

(Fig. 6-15) Cobray M/11 with 8" long suppressor

Cobray M/11 9mm

The gun shown here can be outfitted with an aftermarket stock, scope mount, and other accessories. This suppressor was made fairly large, 8" long by 1 3/4" wide for extra noise reduction, and it works for that very well. Because of the short barrel, the breech blast is not nearly what it is with the Marlin rifle, but if you try firing it while holding a rag over the ejection port the difference is still quite noticeable. A Cobray in .380 caliber would be significantly quieter than the 9mm, and would be about the max I'd want to use indoors in an urban situation that you don't want neighbors to notice.

Every now and then you hear stories of ex-special forces people who make extra money on the weekends taking down drug houses. I don't personally know anyone who has engaged in this type of thing, but it seems to me like a perfectly all-American way to make extra cash and do a public service at the same time. The Cobray in .380, with its high capacity and a good suppressor, would be ideal for this in a situation where you're just kicking the door in. If you needed a concealed weapon, I'd go with the Makarov.

The Makarov has one other advantage by the way too, and that is that you can clear jams a lot more quickly. That's if it even does jam, which would surprise me because I've found it to be extremely reliable when it's not dirty. With the Cobray, you have to drop the magazine out and do some fumbling to clear it. With the Makarov, just a pull of the slide normally does it.

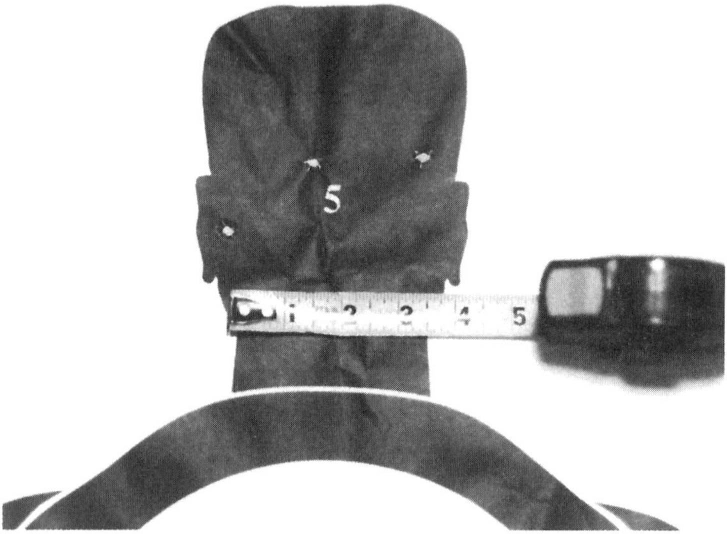

(Fig. 6-16) The Cobray, when accessorized with a stock and scope, is surprisingly accurate with good ammo. This photo shows three consecutive head shots taken at 75 yards using 3D remanufactured ammo.

(Fig. 6-17) .50 BMG caliber rifle with home built suppressor. Here it is, the big kahuna. When I first saw ads for .50 caliber suppressors that suppress down to a .22 noise level, I said "no way." Not only was I wrong, now I've done it myself.

.50 BMG

This size of weapon requires, of course, somewhat heavier materials. The outer tube is made from 2 1/2" muffler pipe with a 3/32" wall thickness. You can get this at an exhaust shop cut to the length you want. The inside diameter of this tube is 2 5/16". This is where a small headache develops. Dorman makes freeze plugs in size 2 1/4", and 2 3/8", but not 2 5/16". So, the solution is to buy them in size 2 1/4", then put them on an anvil or other hard surface, and pound them with a hammer until they become 2 5/16". Just beat the top of the plug flat, and then pound around the edges to expand it until it fits inside the pipe snugly.

The weapon shown here is a LAR Grizzly Big Boar single shot. The muzzle brake for this gun happened to work perfectly for this size muffler pipe. By pounding two sides of the pipe flat to match the shape of the muzzle brake, the pipe can be made to match up nicely to the rear of the muzzle brake for welding. If your rifle has a different muzzle brake that requires creation of a rear endplate, be sure to make it from steel at least 1/4" thick.

For centering, you'll have to use a good straight piece of 3/8" or 7/16" roundbar built up with tape in two spots. Make the

(Fig.6-18) Outer pipe formed and welded to LAR Big Boar muzzle brake. Unwelded muzzle brake shown next to it.

hole in the front centering plug a little larger than the roundbar, so you can see how well it stays centered in the hole as you fit the pipe into place on the muzzle brake. Try to get the outer tube centered on the first try, because if you don't it will be a whole lot harder to adjust than the thinner tubing. Heavy stainless screen strips will be 48" long, same as the width of the roll the screen is cut from.

Using a muzzle brake to build the suppressor is a bit expensive, as they can cost a few hundred dollars by themselves. If you can find a cheaper fitting that fits on your muzzle, you can save some money, but it might not be as effective. Muzzle brakes make excellent blast baffles inside the suppressor.

When welding these heavier materials, you can use a slightly larger 5/64" size welding rod. These rods are also longer, which you will need to reach the first freeze plug baffles at the bottom of the unit.

Shooting Results:

If you shoot this gun in the air in a clear area, you'll be quite pleased with the noise suppression. In most normal shooting environments, it easily passes the no-hearing-protection-required test. On the other hand, you'll also be amazed by how much supersonic crack the big bullet traveling at around 3000 fps generates. The upside of that is how well that noise can mask your location.

Given that I had only a single shot rifle to work with, there wasn't much I could do in the way of torture testing. Still, I did try firing twenty rounds through it as fast as I could load them. It didn't generate excessive heat, and the screen rolls inside held up fine. With a Barrett semi-auto, I'm sure you could get it a whole lot warmer, but it would still hold up fine unless you really put your mind to melting the thing. With a Barrett and a bunch of loaded mags, I'm sure you could do it.

At a hundred yards, the point of impact wasn't changed at all. But as heavy as these barrels and slugs are, it's hard to imagine much of anything changing the point of impact.

As for fun? There is no feeling in the world that is quite the same as quietly blasting flaming holes through 2" thick steel plates with a gun fired from the shoulder. I recommend everyone try it.

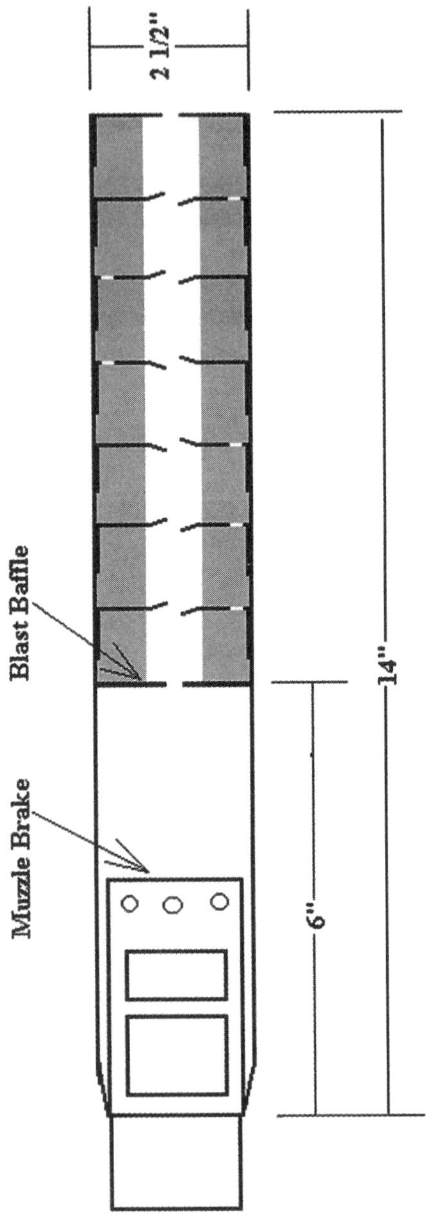

.50 BMG Suppressor

Weapon Light Mounts

CHAPTER 7

For combat type arms, a good weapon light is pretty much indispensable these days. Here are some examples of ways you can construct your suppressors to accept them directly.

(Fig. 7-1) Suppressed Makarov with weapon light mounting rail attached

The M-3 weapon light from Streamlight Inc. is made to mount onto a Weaver rail. So, to mount it onto a suppressor we just need to place a Weaver rail onto the suppressor.

To keep this suppressor from tightening down further as the threads get worn in, this suppressor has been designed with the flash hider sticking out the back end of the unit instead of toward the inside. There is a set screw in the bottom of this flash hider, which seats into a flat recess that comes on the bottom side of the barrel threads of Federal Arms barrels. The hole in the flash hider for this set screw has to be drilled and tapped.

Then, an aftermarket Weaver rail for a 10/22 is cut down to proper length, the M-3 light is placed on, and the rail is positioned on the suppressor. Drill two mounting holes directly in the suppressor, and tap them for the mounting screws, usually 6-48 TPI. Taps of this size are available from Brownell's. Be careful not to tighten the screws down too tight because the suppressor wall material is only 1/16" thick and the threads can be stripped. If that happens don't panic though, you can carefully weld over the hole, grind it flush, and do it over again.

Streamlight Inc. also makes a version called the M-6 which includes a laser sight. If you spend the extra money on that you can have a laser on the suppressed Makarov as well. This would be useful since the sights are obstructed and there are no night sights available for the gun.

Figs. 7-2 and 3 show an alternative way to mount the light. In Fig.7- 2, beneath the pistol is a piece of tubing the same size as the suppressor tube, with a longitudinal section cut out, and the

(Fig.7-2)

circumference slightly opened up with a hammer, so that it will slide over the suppressor body. Also there is the 10/22 Weaver mount. The Weaver mount gets cut down and mounted on this exterior tube.

Fig. 7-3 shows how the unit is assembled, slid onto the suppressor body, and held in place with a hose clamp.

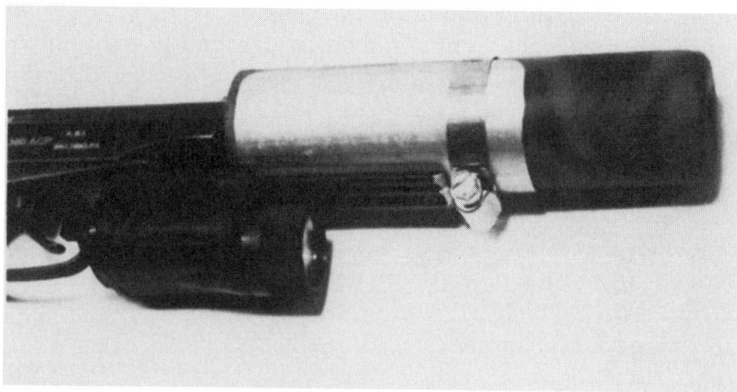

(Fig. 7-3)

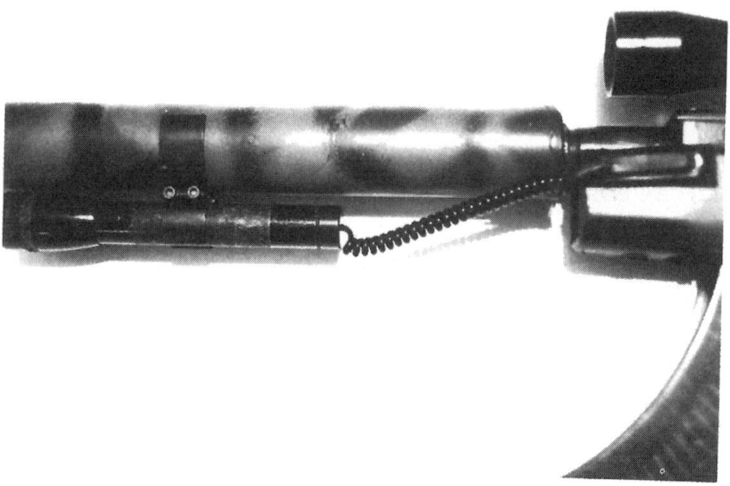

(Fig. 7-4) Mini-Maglite mounted to 10/22 Integral Suppressor

The mount shown in Fig. 7- 4 on the previous page is a Universal Barrel Mount barrel clamp made by Tac-Star. This one is the model UBM-2. The large end has been pounded open and enlarged with a hammer and anvil to fit around the suppressor tube. This suppressor tube made from 1 1/4" conduit was the largest size this clamp would work with.

This time, a small piece of pipe or roundbar with a 3/4" diameter has been tack welded on each end to the bottom of the suppressor tube. The light can then be clamped to the unit as shown. If the location of this mount changes as the barrel threads become worn in, you can break it loose and re-weld it in a different spot.

Sure-Fire makes a couple of weapon light mounts for suppressors. One uses something similar to hose clamps and is adjustable for size from .7" to 1.7". The other is a fixed size clamp made for a 1.6" tube outer diameter. It will fit the 1 1/2" chain link fence pole which has an outer diameter of 1 5/8". These clamps are sold by Shomer-Tec as models MT-3 and MT-23.

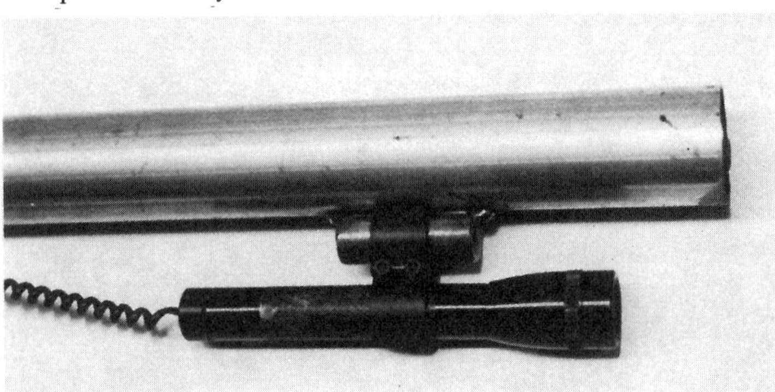

(Fig. 7-5) Another mounting method

Welding Equipment and Procedures

CHAPTER 8

(Fig. 8-1) This is quite possibly one of the most ancient types of stick welding machines still available, but it sufficed nicely for building the suppressors in this book. The basic components of the machine are the ground clamp, welding rod holder (stinger), and power level selector switch.

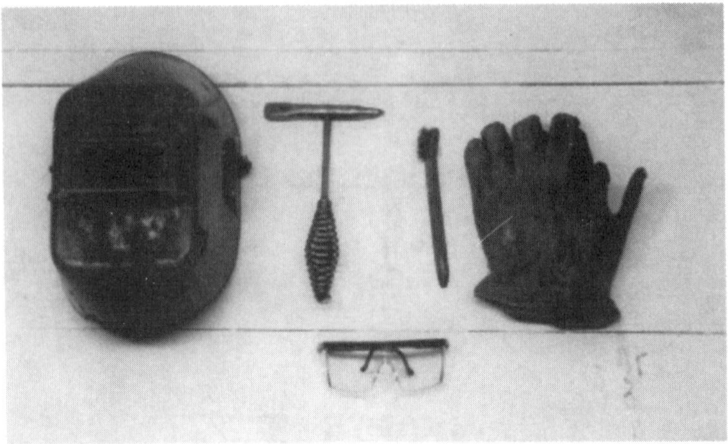

(Fig. 8-2) Shown here are the basic items you will be needing for welding. 1. Welding hood, with Shade 10 or 11 dark lens. 2. Slag hammer, also known as a chipping hammer. 3. Small wire brush. 4. Leather gloves. 5. Safety glasses.

Welding is a skill that can take weeks of practice and instruction to attain basic proficiency. Fortunately, the skill required to build suppressors is not especially great.

The first step will be selecting the welding machine you will be working with. There are three basic types to consider. The first is a stick welding machine. Stick welders can be found used, often very inexpensively, or they can be rented. In the long run, you're probably better off buying one, especially if you will be needing welding practice. You can buy a beat-up used machine for about the cost of renting one for two or three weekends. If you have little or no welding experience, a stick machine should be your first choice, as they require the least expertise. Virtually any machine that still makes noise when you turn it on will suffice. There are even machines that run on 110v power available that are plenty powerful for the types of welding rod used for suppressor building. If possible, select a machine that has a rheostat type power adjustment, rather than one that only switches between different pre-set power levels, such as the one shown.

This will allow you to set your power level more precisely, which is very helpful, as welding thicker materials requires just slightly more power than thin materials. Stick welding machines come in AC and DC types. The DC welders run smoother and weld more easily, but AC machines are less expensive and work well enough for these projects.

The welding rod size you will be needing is 1/16" diameter, in a type suitable for welding mild steel. There are only two types of rod that come in that size that will work well for building suppressors with the materials we use. One of these is called "Multi Purpose Welding Rod", and is made by Century Manufacturing. The other is called 309. 309 welding rod is actually made for welding stainless steel to black carbon steel. It works well for welding just carbon steel though, and it tends to run a little smoother and a lot cleaner than the "Multi-Purpose" type. There is a lot less spatter and smoke created by 309, and the slag is a lot easier to chip out. While both types will result in sufficiently strong welds, and require about the same skill, the 309 is preferable to work with. If you are using slightly heavier materials to build the suppressor such as with the .50 BMG unit, you can use size 5/64" welding rod. In this size there is another type of rod available, called 6013. It makes a nice smooth weld and is recommended when 5/64" rod can be used. Unfortunately 6013 isn't made in 1/16".

The second type of machine is the wire feed MIG welding machine. This is a wire feed welder that uses gas coverage from the welding nozzle to protect the weld process from oxygen, which would otherwise burn around the weld, ruining it. MIG welding is more complicated and difficult to learn, as you have to become familiar with more machine settings in order to make it run smoothly, as well as learning better technique. It's also about impossible to get the nozzle down into the tube where you will be doing your welding.

There are a couple of other types of wire feed welders. One is called Dual Shield. This type uses a combination of gas coverage and a flux coated wire to shield away oxygen. This type of welder generates too much heat however, and would vaporize most of the materials you will be working with, and is thus unsuitable.

The other type of wire feed welder is called Innershield. This type uses only a flux-cored wire as an oxygen shield. The smoke it creates is what acts as the oxygen shield. This welding process is crude, messy, and will suffice only if you have welding experience with it.

For instruction purposes, we will be concentrating on stick welding. Anyone who has experience with a MIG machine doesn't need further instruction from me, and beginners are far better off starting off with a stick machine because of its simplicity. It is also by far the least expensive process in terms of equipment and supplies needed.

Striking An Arc

The first step in stick welding is being able to set your machine to the right power level, and then to strike an arc and keep it going. Power settings are fairly simple. 1/16" welding rod usually uses the lowest setting on a stick machine, or very close to it, since it is the smallest size generally available.

Begin with a scrap piece of steel. Place it flat on your workbench so you are welding across a flat surface. Connect the ground clamp to the material being welded, and place a welding rod in the welding stinger. Place the tip of the rod close to the point you want to start welding on, then flip your hood down. Lightly scratch the rod across the metal as though you were striking a match, then draw it away just slightly. Give it just a moment for the end of the rod to heat up, then you can push the rod down to lightly contact the metal. Metal will then flow from the rod into the metal being welded.

The condition of the flux on the end of the rod makes a huge difference in how easily it will strike an arc. Broken away flux will not work and you will need a new rod. If you are properly trying to strike an arc inside a suppressor tube but can't get one going, and the flux looks okay, then take the rod out and strike an arc on a piece of scrap metal just until a smooth arc is going. Then stop welding on the scrap, and with the tip still hot, move the rod back to where you were trying to weld before and you should be able to get an arc going easily then. If not, then gradually turn the power up until it does work.

Basic Technique

Now we will begin to practice the basic techniques you will need to make a good weld. After you have struck an arc on the flat piece of steel, push it into the metal lightly, and begin to draw it toward you slowly, in order to form a smooth bead of weld. Watch the puddle of molten metal closely, and watch how the bead forms. As you draw it toward you, concentrate on forming a smooth, even, and very consistent bead. A good bead has straight, even edges, does not waver back and forth, and does not have lumps and thin spots from speeding up and slowing down. After welding the bead, use the slag hammer to break off the slag coating, and the wire brush to clean it up and see how it looks.

If you are new to welding, plan on spending some time at this. It takes some time to get the hang of it.

(Fig. 8-3) Here are some examples of practice beads on a piece of scrap steel. Note that the third one is the straightest and most consistent.

Fillet Welding

A fillet weld is a weld joining two pieces of metal that are joined at right angles to one another. Refer to the drawing below. To practice this weld, place two pieces of flat scrap steel together as shown. Use the welder to tack them together on each end.

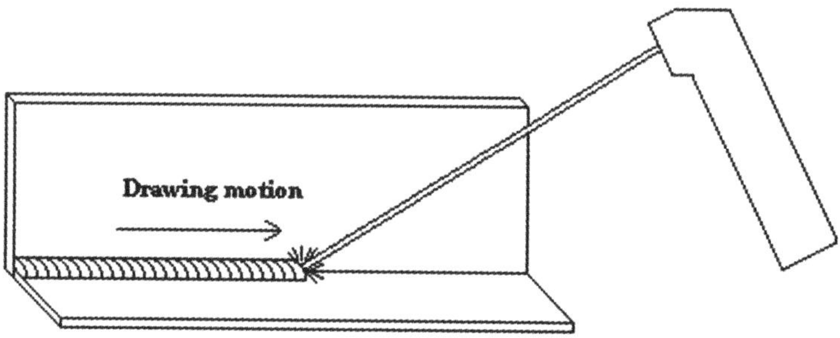

First strike an arc on the left end of the plates, and move the rod into the corner. As you draw the rod to the right, you must also move it upward and downward, smoothly and consistently, to wash the metal into both plates. What you are looking for is to see the arc penetrate into both plates as you oscillate the rod up and down. Watch the puddle of weld closely and make sure that the same puddle continues along, being washed into both plates, rather than just making a zigzag pattern which will leave gaps. This is how you get good fusion between the welding rod metal and the two pieces of material being welded.

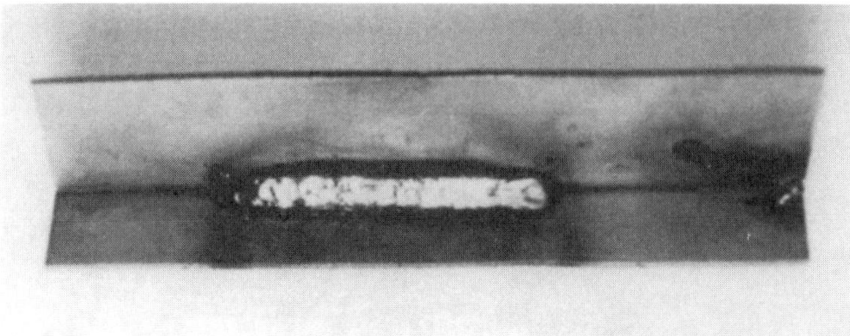

(Fig. 8-4) Example of a basic fillet weld. Note that the weld is straight and consistent, and solidly fused to both pieces of the material.

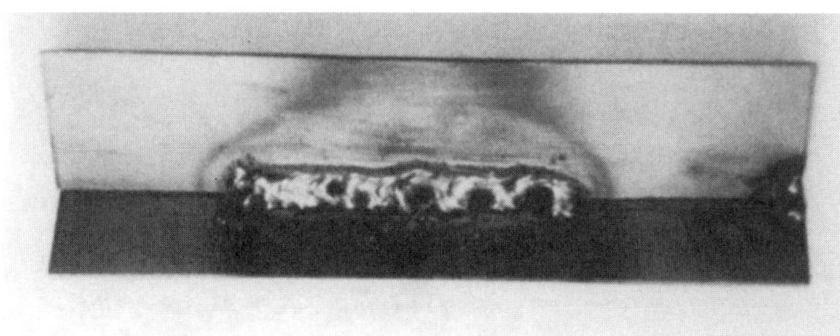

(Fig. 8-5) Example of a bad fillet weld. It is inconsistent, and insufficiently fused to both pieces of material. Possible causes include poor technique, wrong type of welding rod, and machine power setting too low.

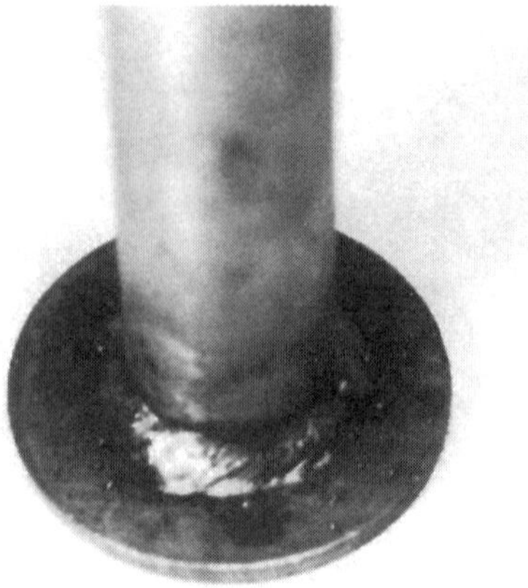

(Fig. 8-6) This is a fillet weld as applied to a centerpipe and an endplate. Technique is the same, you're just welding in a circle.

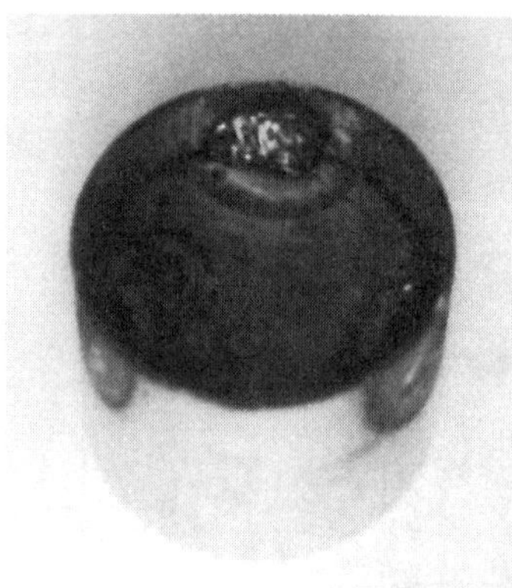

(Fig. 8-7) Here is a fillet tack weld as applied to freeze plug baffles. It is short, but solidly fused to both baffle and outer pipe. These materials are somewhat difficult to get fused on the first try, so you may need to chip out all the slag from the first attempt and re-weld over the same spot again. To do this without burning through the materials, it often helps to use a lower power setting for the first attempt, just to get some metal deposited onto both pieces. Then, turn up the power to fuse the two pieces. The thickened metal will not burn through as easily.

Vertical Welding

There are two basic motions for vertical welding. One is called uphand, the other downhand. Uphand welding is used for thicker materials. After striking an arc at the bottom of the weld, the rod is oscillated between the two pieces of material in such a way that the weld metal is deposited on top of itself as the weld progresses upward. This method generates more heat and penetration, and more weld metal deposited into the material. Uphand welding is not for thin materials, as it will burn through them much more easily.

For lighter materials, such as when welding endplates to outer pipes, the downhand method is used. To make this weld, begin at the very top of the outer pipe, strike the arc, and begin welding toward you and down. Continue until you reach the midpoint. You will need to move fairly quickly to avoid burning through the outer pipe. There is no need to oscillate the rod back and forth for this weld either, as the materials are thin and weld quickly together. When you reach the midpoint, stop and rotate the pipe in the vise, and continue the weld.

(Fig. 8-8) Here is an example of a downhand welded outer pipe and endplate.

Splicing Pipe

Being able to splice pipe is necessary for welding extension pipes onto flash suppressors to create center pipes for suppressors that use that type of design. Depending on the thickness of the pipes being welded, you may be using downhand welding, uphand welding, or a combination of the two.

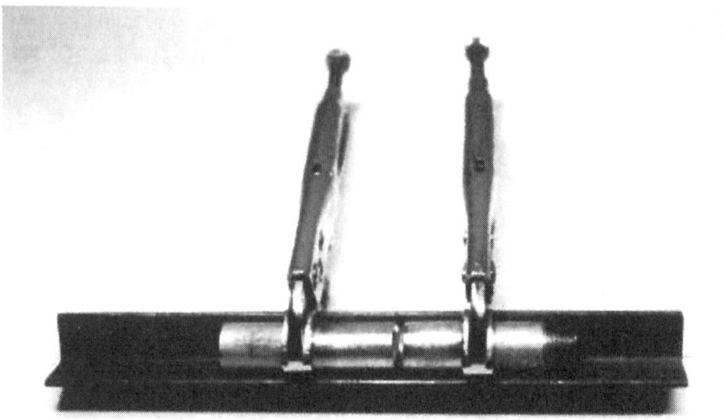

(Fig.8-9) Two pieces of pipe prepared for welding.

In Fig. 8-9, note that a bevel has been ground onto each piece which is then to be filled with weld. The two pieces are clamped into a piece of angle iron to hold them in alignment. If one piece of pipe is slightly larger than the other, you can place a piece of paper, cloth, or something else under the smaller pipe to raise it up and center it in alignment with the larger one.

These pipes can then be welded downhand similar to the previous outer pipe. To make a thicker, stronger weld, they can also be welded uphand, as described previously. Since this method creates more heat and weld penetration, the weld may burn through into the inside of the pipe. If you encounter this, you can start the weld by making a thin downhand root pass around the pipe, and then welding over the top of that with an uphand weld.

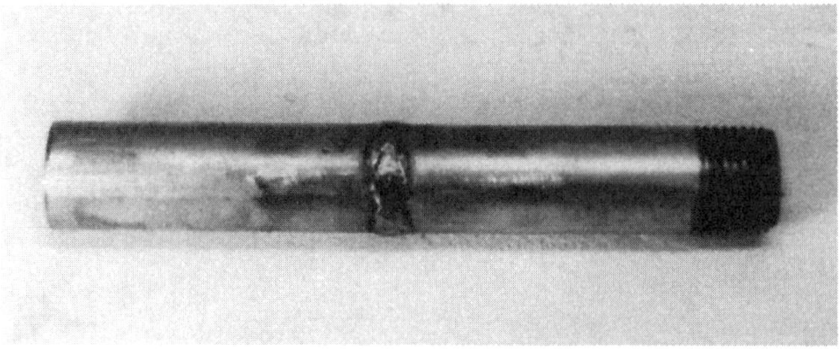

(Fig. 8-10) Two pieces of pipe spliced and welded.

Troubleshooting

There are certain problems that are associated with welding. One of them is called "spatter." These are little beads of metal that spray away from the weld and stick to metal wherever they land. They are not a problem unless they land on the threads inside the unit and stick. Then they can make threading the unit on difficult, and cause barrel thread damage. The solution is to use a threading tap to clean up the threads if this happens. Spatter on threads can generally be prevented by covering them with a small piece of metal, or protecting them with a plug of some type, even a round piece of wood pushed into them.

"Heat shrinkage" is another one to look out for. When welded metal is heated red hot, it of course expands. As it cools, it does not return to its previous size, but actually contracts slightly smaller than before. This is why when you weld directly around the threads in your flash hider, the shrinkage can make it difficult or impossible to thread the unit onto your barrel. When that happens, you must again use a tap to clean up the threads.

"Porosity" is a condition where the weld is full of little holes, like bubbles. This can result in a weaker weld, more prone to cracking. Porosity is usually attributed to moisture on the steel being welded, moisture in the welding rod, or corrosion on the steel. Rod and steel can be heated to 150 to 200 degrees in an oven to remove all traces of moisture. Corrosion must be thoroughly cleaned off with a wire brush.

How to Build Military Grade Suppressors

Source Directory

Most of these companies have websites that list all their products. Call for their website address.

AMMO

3-D Remanufactured Ammunition
P.O. Box J
Doniphan, NE 68832
(402) 845-2285

Carries good quality inexpensive ammo that runs at subsonic speeds in all pistol calibers including 9mm (147 grn.) and .380 acp. Call for brochure of all their loads.

Engel Ballistic Research
544A Alum Creek Road
Smithville, TX 78957
(512) 360-5327

Sells top-of-the-line subsonic sniping ammo in .223, .308, and 7.62x39 to law enforcement or FFL holders. The 7.62x39 will cycle the weapons, and uses a proprietary 220 grain bullet design that is stable with rifling twist rates as slow as 1:12. They are at this time working on a .308 version that will do the same thing.

SRT Arms
1739 East Broadway Rd. #1-161
Tempe, AZ 85282
(480) 967-6218

Supplier to the public of Engel Ballistic Research ammo.

COR-BON Bullet Co.
1311 Industry Road
Sturgis, SD 57785
(605)-347-4544

Sells .300 Whisper ammo in three different loads, including 220 grain subsonic.

Champion Shooters Supply
P.O. Box 303
42 N. High St.
New Albany, OH 43054
(614) 855-1603

Sells just about every kind of .22 ammo in existence, including Lapua Scoremax.

Gun City Gun & Cartridge Co.
309 Hwy. 400A
Dahlonega, GA 30533
(706) 867-8313

Another supplier of all types of .22 ammo.

Extreme Shock Ammunition
Rt. 2 Box 304-K
Clintwood, Virginia 24228
(540) 928-6772

Makes highly advanced (and highly expensive) subsonic frangible ammunition in .223 caliber, and is working on other calibers. They have the first subsonic .223 round that will cycle an AR-15 and stabilize the bullet in a 1x7 twist barrel.

HJ Ballistic Research
1127 Old Centre Hwy.
Piedmont, AL 36272-6429
(256) 447-1987

PARTS AND ACCESSORIES

Federal Arms Corporation
7928 University Avenue
Fridley, Minnesota 55432
(612) 780-8780

Supplier of after-market threaded barrels. Also carries hard to find flash suppressors, including Marlin Camp Carbine, twist-on type for SKS, and Makarov. Many other accessories. This is a must-have catalog.

RPB Industries
P.O. Box 367
Avondale Estates, GA 30002
(404) 297-0907

Cobray M11/9 machine pistol accessories and parts. Also has Ruger Mark II .22 pistol threaded barrels.

Brownell's, Inc.
200 South Front Street
Montezuma, IA 50171
(515) 623-4000

Aerosol Baking Lacquer, in matte black and other camouflage colors. Also has threading taps in 1/2 x 28 and other sizes.

K.G. Industries
537 Louis Drive
Newbury Park, CA
(800) 348-9558

Supplier of Gun-Kote coating product. Gun-Kote is available in many colors, in either gloss or flat.

Delta Press
215 South Washington Ave. Dept. V48
El Dorado, AR 71730
(800) 852-4445

Sells the book "Ultimate Ruger 10/22" by Mark White, among hundreds of other good titles.

Delta Force Catalog
215 South Washington Ave. Dept. 23
El Dorado, AR 71730
(800) 852-4445

Sells a variety of weapon lights and mounts. Also has a huge variety of gun accessories and tactical equipment.

Numrich Gun Parts Corporation
226 Williams Lane
West Hurley, NY 12491
(845) 679-2417

Sells rear-extended AR-15 flash hiders. Item # 555660.

McNichols Co.
P.O. Box 30300
Tampa, FL 33609-1007
(800)-237-3820

Supplier of heavy stainless screen. The two mesh sizes you'll need are:

1. **16 mesh, .018 wire thickness**
2. **12 mesh, .028 wire thickness.**

The mesh size is the number of wires per inch. The lighter 16 mesh is for pistol and other light calibers. The heavier 12 mesh is for rifle calibers. This screen comes on 4-foot wide rolls, and they can cut it from the roll to the length you need.

They also have the 12 mesh, .028 wire thickness in plain carbon steel, and also a size 14 mesh, .020 wire thickness in

carbon steel. Carbon steel mesh is cheaper, but will rust, even if coated. Ask for woven mesh, not welded mesh.

Muscle Products Corporation
112 Fennell Drive
Butler, PA 16002
(724) 283-0567

Sells PL-10 Lithium Grease made for use inside suppressors to increase suppression effect similar to how water does.

SSK Industries
590 Woodview Lane
Wintersville, OH 43953
(740) 264-0176

Sells AR-15 upper receivers in .300 Whisper caliber made to function with Cor-Bon ammo. Also sells dies for converting .221 Fireball brass to .300 Whisper brass, if you own one of their AR uppers.

LBC Autoparts Warehouse
135 South Fifth Street
Newport, PA 170-1009
(800) 382-1322

Sells Dorman freeze plugs in all sizes. This is a preferred brand. The type you want is always steel deep cup.

WEAPON LIGHTS

Shomer-Tec
P.O. Box 28070
Bellingham, WA 98228
(360) 733-6214

Sells weapon light mounts for suppressor tubes, among many other law enforcement items.

Streamlight, Inc.
30 Eagleville Road
Eagleville, Pa. 19403
(800) 523-7488

Manufacturer of the M-3 tactical light, and M-6 tactical light/ laser.

Tac-Star
475 Smith St.
Middletown, CT 06457
(800) 423-9704

Makes the Universal Barrel Mount, a simple weapon light clamp in several sizes. Also sells inexpensive weapon lights and lasers.

WELDING

Century Mfg. Co.
9231 Penn Avenue South
Minneapolis, MN 55431
(800) 328-2921 or (612) 884-3211

Carries Multi-Purpose Welding Rod, also carries several welding machines. (Model 84100 recommended)

Harris Welco
1051 York Rd.
Kings Mtn., NC 28086
(704) 739-6421

Carries 1/16" 309 welding rod.

Defend and protect the Constitution or just settle your scores with that noisy neighbor dog!

Notes

Notes

Notes

Notes

Notes

Notes

Other Books Available From Desert Publications

081 Crossbows/ From 35 Years With the Weapon $15.95
206 CIA Field Expedient Preparation of Black Powder $19.95
221 Evaluation of Improvised Shaped Charges $19.95
411 Homeopathic First Aid $14.95
454 Survival Childbirth $15.95
C-002 How To Open A Swiss Bank Account $9.95
C-011 Defending Your Retreat $9.95
C-020 Methods of Long Term Storage $9.95
C-023 Federal Firearms Laws $11.95
C-024 Select Fire Uzi Modification Manual $13.95
C-025 How to Build Silencers $6.95
C-027 Firearms Silencers Volume 1 $13.95
C-029 M1 Carbine Arsenal History $7.95
C-032 The Silencer Cookbook $9.95
C-038 How To Build A Beer Can Morter $4.95
C-040 Criminal Use of False ID $14.95
C-045 Improvised Weapons of American Underground $14.95
C-048 Catalog of Military Suppliers $18.95
C-056 Navy Seal Manual $29.95
C-065 Poor Man's James Bond Vol 1 $34.95
C-113 Submachine Gun Designers Handbook $19.95
C-114 AR-15, M16 and M16A1 5.56mm Rifles $13.95
C-117 Ruger Carbine Cookbook $13.95
C-120 Shootout II $14.95
C-122 AK-47 Assault Rifle $15.95
C-123 Combat Loads for Sniper Rifles $18.95
C-125 Browning Hi-Power Pistols $13.95
C-126 UZI Submachine Gun $13.95
C-127 FullAuto Vol 1 Ar-15 Modification Manual $11.95
C-128 Full Auto Vol 2 Uzi Mod Manual $11.95
C-130 Full Auto Vol 4 Thompson SMG $11.95
C-131 FullAuto Vol 5 M1 Carbine to M2 $11.95
C-135 M-14 Rifle, The $12.95
C-136 Fighting Garand Manual $13.95
C-137 Ranger Training & Operations $19.95
C-138 Emergency War Surgery $29.95
C-140 Emergency Medical Care/Disaster $14.95
C-143 Survival Medicine $15.95
C-150 Improvised Munitions Black Book Vol 1 $17.95
C-151 Improvised Munitions Black Book Vol 2 $17.95
C-152 Impro. Munitions Black Book Vol 3 $29.95
C-155 Fighting Back on the Job $14.95
C-162 Colt .45 Auto Pistol Manual $13.95
C-163 Survival Evasion & Escape $19.95
C-165 USMC Sniping $19.95
C-175 Beat the Box $7.95
C-197 How to Make Disposable Silencers Vol. 1 $19.95
C-198 Expedient Hand Grenades $21.95
C-199 U. S. Army Counterterrorism Training Manual $19.95
C-203 Explosives and Propellants $19.95
C-204 Improvised Munitions/Ammonium Nitrate $19.95
C-205 Two Component High Explosives Mixtures $19.95
C-207 CIA Improvised Sabotage Devices $19.95
C-208 Spcial Forces Demolitions Training Handbook $19.95
C-210 Guerilla Warfare $17.95
C-211 FN-FAL Auto Rifles $21.95
C-218 Professional Homemade Cherry Bomb $15.95
C-219 Improvised Shaped Charges $19.95
C-220 Improvised Explosives, Use In Detonation Devices $19.95
C-227 Springfield '03 Rifle Manual $15.95
C-229 Lock Picking Simplified $11.95
C-230 How to Fit Keys by Impressioning $9.95
C-231 Lockout -Techniques of Forced Entr $15.95
C-235 How to Open Handcuffs Without Keys $13.95
C-244 Camouflage $14.95
C-274 M14 and M14A1Rifles and Rifle Marksmanship $23.95
C-284 Thompson Submachine Guns $19.95
C-289 M1 Carbine Owners Manual $13.95
C-290 Clandestine Ops Man/Central America $14.95
C-324 How to Make Disposable Silencers Vol. 2 $19.95
C-337 Sten MKII SMG Construction Manual $24.95
C-356 Map Reading and Land Navigation $19.95
C-365 Survival Gunsmithing $14.95
C-372 Poor Man's James Bond Vol 2 $34.95
C-386 Self-Defense Requires No Apology $15.95
C-414 Select Fire 10/22 $15.95
C-415 Construction Secret Hiding Places $15.95
C-416 Brown's Alcohol Motor Fuel Cookbook $24.95
C-417 Canteen Cup Cookery $9.95
C-418 Improvised Batteries/Detonation Devices $19.95
C-421 Walther P-38 Pistol Manual $13.95
C-422 P-08 Parabellum Luger Auto Pistol $14.95
C-423 Sten Submachine Gun, The $11.95
C-427 How to Train a Guard Dog $16.95
C-429 The Anarchist Handbook Vol. 1 $14.95
C-430 Beretta - 9MM M9 $15.95
C-432 Mercenary Operations Manual $14.95
C-439 Improvised Lock Picks $17.95
C-443 How to Disappear Completely & Never be Found $19.95
C-455 Modern Day Ninjutsu $17.95
C-512 FullAuto Vol 8 M14A1 & Mini 14 $11.95
C-518 L.A.W. Rocket System $14.95
C-528 With British Snipers, To the Reich $29.95
C-539 Dental Specialist $59.95
C-558 Food. Fur and Survival $17.95
C-561 The Squeaky Wheel $17.95
C-562 Survival Chemist Book $17.95
C-564 Medical Specialist $59.95
C-571 U. S. Marine Corps Scout/Sniper Training Manual $27.95
C-610 Aunt Bessie's Wood Stove Cookbook $9.95
C-622 Defensive Shotgun $14.95
C-625 MAC-10 Cookbook $14.95
C-627 Brown's Lawsuit Cookbook $26.95
C-628 Can You Survive $16.95
C-633 Hand to Hand Combat $11.95
C-634 USMC Hand to Hand Combat $9.95
C-635 Hand to Hand Combat by D'Eliscue $11.95
C-636 Prisons Bloody Iron $16.95
C-639 Police Karate $17.95
C-646 Science of Revolutionary Warfare $16.95
C-668 Survival Shooting for Women $14.95
C-680 Infantry Scouting, Patrol. & Sniping $16.95
C-694 CIA Field Expedient Incendiary Manual $19.95
C-761 How to Lose Your X-Wife Forever $24.95
C-763 Vigilante Handbook $19.95
C-775 Alcohol Distillers Handbook $27.95
C-776 How to Build a Junkyard Still $15.95
C-777 Engineer Explosives of WWI $19.95
C-794 US Marine Corps Essential Subjects $19.95
C-815 US Marine Bayonet Training $11.95
C-818 The Poisoner's Handbook $29.95
C-823 Cold Weather Survival $14.95
C-829 The Anachist Handbook Vol. 2 $14.95
C-840 Freedom Road $21.95
C-889 Napoleon's Maxims of War $11.95
C-890 Company Officers HB of Ger. Army $19.95
C-891 German Infantry Weapons Vol 1 $22.95
C-896 Op. Man. 7.62mm M24 Sniper Weapon $13.95
C-899 US Army Bayonet Training $11.95
C-9002 German MG-34 Machinegun Manual $19.95
C-9012 Firearm Silencers Vol 3 $18.95
C-9013 HK Assault Rifle Systems $27.95
C-9047 SKS Type of Carbines, The $17.95
C-9048 Rough Riders, The $24.95
C-9052 Private Weaponeer, The $17.95
C-9057 Keys To Understanding Tubular Locks $11.95
C-9060 The Anarchist Handbook Vol. 3 $14.95
C-9080 Leadership Hanbook of Small Unit Ops $14.95
C-9081 Dirty Fighting $13.95
C-9082 Lasers & Night Vision Devices $29.95
C-9083 Water Survival Training $8.95
C-9084 Assorted Nasties $29.95
C-9085 Ruger P-85 Family of Handguns $19.95
C-9101 Guide to Germ Warfare $17.95
C-9102 CIA Field Expedient Method for Explosives Preparation $19.95
C-9119 Surviving Global Slavery $16.95
C-9120 Military Knife Fighting $14.95
C-9127 H&R Reising Submachine Gun Manual $19.95
C-9131 Live to Spend It $29.95
C-9134 Military Ground Rappelling Techniques $15.95
C-9135 Smith & Wesson Autos $19.95
C-9137 Caching Techniques of U.S. Army Special Forces $14.95
C-9138 USMC Battle Skills Training Manual $49.95
C-9139 Survival Armory $27.95
C-9156 Concealed Carry Made Easy $17.95
C-9164 The L'il M-1, The .30 Cal. M-1 Carbine $17.95
C-9170 Urban Combat $27.95
C-9178 Apocalypse Tomorrow $17.95
C-9198 MP40 Machinegun Manual $16.95
C-9199 Clear Your Record & Own a Gun $24.95
C-9200 Sig's Handguns Manual $19.95
C-9210 Sniper Training $24.95
C-9212 Poor Man's Sniper Rifle $19.95
C-9221 The Official Makarov Pistol Manual $14.95
C-9224 The Butane Lighter Hand Grenade $14.95
C-9228 Unarmed Against the Knife $14.95
C-9229 Black Book of Booby Traps $29.95
C-9239 Militia Battle Manual $22.95
C-9247 Black Book of Revenge $15.95
C-9253 Bankruptcy, Credit and You $18.95
C-9255 The Sicilian Blade $15.95
C-9259 How to Build Practical Firearm Silencers $18.95
C-9262 Glock's Handguns $19.95
C-9263 Heckler and Koch's Handguns $19.95
C-9264 The Poor Man's Ray Gun $11.95
C-9265 The Poor Man's R. P. G. $24.95
C-9269 David's Tool Kit $27.95
C-9279 Credit Improvement & Protection Handbook $19.95
C-9281 Build Your Own AR-15 $19.95
C-9282 How to Live Safely in a Dangerous World $19.95
C-9293 Bllack Book of Arson $24.95
C-9322 How to Build Flash/Stun Grenades $19.95
C-9342 Entrapment $19.95
C-9359 The FN-FAL Rifle Et Al. $21.95
C-9367 The Mental Edge, Revised $17.95
C-9373 Survival Bible by Duncan Long $59.95
C-9405 The Internet Predator $17.95
C-9406 The Criminal Defendants Bible $49.95
C-9430 Protect Your Assets $24.95
C-9432 How To Build Military Grade Suppressors $21.95
C-9441 Mossberg Shotguns $24.95
C-9478 Poor Man's Primer Manual, The $24.95
C-9546 Agents HB of Black Bag Ops. $14.95
C-9556 Improvised Rocket Motors $19.95
C-9612 Homestead Carpentry $14.95
C-9633 Getaway $13.95

PRICES SUBJECT TO CHANGE WITHOUT NOTICE

Desert Publications

870-862-3811

215 S. Washington Ave Dept. - BK -195

El Dorado, AR 71730 USA

Shipping & Handling
1 item $6.95 - 2 or more
$11.95